LAND-CHANGE SCIENCE IN THE TROPICS: CHANGING AGRICULTURAL LANDSCAPES

LAND-CHANGE SCIENCE IN THE TROPICS: CHANGING AGRICULTURAL LANDSCAPES

Edited by

Andrew Millington
Texas A & M University
College Station, TX, USA

and

Wendy Jepson
Texas A & M University
College Station, TX, USA

Editors
Andrew C. Millington
Texas A & M University
College Station, TX, USA

Wendy Jepson
Texas A & M University
College Station, TX, USA

ISBN: 978-0-387-78863-0 e-ISBN: 978-0-387-78864-7

Library of Congress Control Number: 2008923743

Printed on acid-free paper

9 8 7 6 5 4 3 2 1

springer.com

TABLE OF CONTENTS

Foreword

Land in the tropics and sub-tropics occupy approximately 40 percent of the Earth's surface and is currently home to a a large portion of the world's population. This book provides a detailed scientific account of the current state and condition of land change in the Tropics. The main themes of tropical land-change science include not only extensification and intensification, but also diversification and competition for land; resilience of land systems; the multiple roles of institutions, markets, societies, and individuals; and the effects of decisions made at manifold spatial, temporal, and organizational scales in influencing land change. These themes together with issues such as frontier settlement, dynamics of plant invasions and other changes in environmental quality associated with alternate land uses clearly demonstrate the importance of an integrated and interdisciplinary understanding of socio-economic and human systems as well as environmental systems. This book takes such a coupled approach to human and natural systems and investigates land change as an exemplar of the fundamental interdependence of society, economy, and environment.

Development of methodologies required for achieving a more integrated and interdisciplinary understanding of land change in coupled natural and human systems are an important effort in the international land-change science community. This book addresses and explores many of these methodologies, providing detailed case studies that demonstrate the importance of strong methodologies. Specifically, the examples include linking social science methods with environmental science methods; integrating remote sensing with household and community surveys; using spatial analysis; and basing analysis on integration of a variety of disciplines including economics, sociology, anthropology, geography, and landscape ecology, as well as on deep knowledge of individual land uses such as agriculture and forestry.

The examples and analysis also provide a detailed examination of many of the generic themes of land-change science and the diverse land-use systems that are found in other places on the surface of the Earth. Many of these generic themes are obviously scientific. For example, improved understanding of the resilience of long-settled and natural landscapes to land management practices, including technological change, as well as to change in land use and other environmental changes are important for advancing understanding of human-environment relationships. Scientific study of stasis in landscapes allows process based understanding of change to be coupled with place-based environmental and social histories to understand long-term evolution and coadaptation between environments and societies. Improved understanding of agents and drivers of change, and other processes by which change occurs, also advance land-change science. The generic themes are also, however, of a strongly applied nature, with direct relevance to policy and practice at local, regional, national, and global scales. Understanding the resilience, economics, social importance, and environmental challenges and opportunities can lead to the development of policies and land management plans that contribute to the sustainability of communities and economies. These scientific and applied themes are of increasing importance in the context of both human and environmental dimensions of global change and as issues of food security, energy security, environmental security, global trade, and social and environmental justice attract political, economic and societal attention.

This book contributes to a growing international effort to understand the nature, trajectory, and consequences of land change. International programmes such as the International Geosphere-Biosphere Programme/International Human Dimensions Programme *Global Land Project* network the efforts of an international community of land-change scientists to develop land-change science as a basis for knowledge and policies that improve the sustainability of the Earth's land systems. This book makes a valuable addition to this science and will be of interest to all in this global network.

Richard Aspinall, Macaulay Institute, Aberdeen, Scotland

CONTRIBUTORS

Keila Aires was born in the State of Rondônia, Brazil, and is currently pursuing a masters degree in environmental economics and policy at Duke University, Durham, North Carolina. She is researching financial mechanisms aimed at reducing deforestation and fostering local development in Brazil. Previously she worked in Rondônia for the Amazonian Large Scale Biosphere-Atmosphere Experiment, and has also assisted various local non-profit organizations to establish their environmental management systems and fundraising activities. Her professional interests include environmental economics and policy, corporate environmental social responsibility, deforestation and climate change, and sustainable development

Andrew Bradley has a B.Sc. from University of Hull, and M.Sc. and Ph.D. degrees from the University of Leicester. His Ph.D. research focused on land-use change in Chapare, Bolivia. He has held positions as a Research Associate in the School of Geography and Environmental Science at the University of Birmingham, as a Temporary Lecturer in the Geography Department at the University of Leicester, and as a Post-doctoral Research Fellow at the University of Reading. He is currently a Remote Sensing Specialist at the Centre of Ecology and Hydrology in Cambridgeshire, U.K.

Thomas J. Bassett is Professor of Geography at the University of Illinois, Urbana-Champaign. He received his Ph.D. and M.A. degrees in Geography from the University of California at Berkeley. He holds B.A. degrees in English and French from Tufts University. His current research focuses on the geography of hunger and the impact of neo-liberalism on natural resource control, access and management in West Africa. He is the author of *The Peasant Cotton Revolution in West Africa, Côte d'Ivoire, 1890-1995* (Cambridge 2001), and the co-author of *African Savannas: Global Narratives and Local Knowledge of Environmental Change* (James Currey & Heinemann, 2003); *Political Ecology: An Integrative Approach to Geography and Environment-Development Studies* (Guilford, 2003), and *Nature as Local Heritage in Africa* (International African Institute, 2007).

Molly E. Brown is a Research Scientist at Science Systems and Applications, Inc., working at the NASA Goddard Space Flight Center in Greenbelt, Maryland. She obtained her doctoral and masters degrees in Geography from the University of Maryland, College Park, and also has a B.S. in Biology from Tufts University. Her current work focuses on the interface between social and physical sciences, particularly in early warning systems and climate change research in Africa. She has received several grants from NASA to augment existing decision support systems with remote sensing data, enabling enhanced understanding of the impact of climatic extremes on food security.

J. Christopher Brown is an Associate Professor of Geography and the Environmental Studies Program at the University of Kansas. His Ph.D. is in Geography from the University of California, Los Angeles. Earlier degrees include a B.A. in Biology and an M.A. in Latin American Studies from the University of Kansas. His research centers on the human and environmental dynamics of commodity production in Latin America. His most active project, at present, deals with the growth of mechanized annual cropping in the Brazilian Amazon. This work, often involving collaborations with satellite remote sensing specialists in the US and Brazil, specifically aims to address debates about how to meet the often-contradictory goals of conservation and development.

Laura Ediger completed a Ph.D in Ecology from the University of Georgia in 2006, and continues to work on issues related to natural resource management in China and Southeast Asia. Her interests include landscape ecology, household economic strategies, migration, and large-scale changes in land use. She is currently a Research Fellow at IGSNRR in Beijing.

Jeffrey Edmeades is a social demographer at the International Center for Research on Women. He obtained his doctorate in sociology from the University of North Carolina at Chapel Hill, and also has degrees from the University of Waikato (M.A. in Demography and B.A. in Geography). His current research focuses on the determinants of use of contraception and abortion in India, the effect of increased contraceptive use on women's empowerment, and the effect of gender inequality on economic growth. His has also worked on related issues in New Zealand and Thailand.

Barbara Entwisle is Kenan Professor of Sociology and Director of the Carolina Population Center at the University of North Carolina at Chapel Hill and is currently President of the Population Association of America (PAA). She is a social demographer interested in population dynamics and demographic responses to social change. Her projects have addressed fertility, family planning program evaluation, neighborhood and community contexts, household change and social networks, and migration, household formation, and land-use change. She has particular expertise in the measurement and analysis of community and neighborhood characteristics and the integration of surveys with other forms of data, including administrative records, qualitative observations, remotely sensed data, and other types of spatial data.

Helmut Geist holds a Ph.D. in economic geography from the University of Würzburg, Germany. He is Professor of Human Geography at the University of Aberdeen, U.K, and was Executive Director of the Land-Use/Cover Change (LUCC) project of IGBP and IHDP at the University of Louvain-la-Neuve, Belgium from 2000 to 2005. His interests include human-environment interactions, agroindustrial commodity chains, and political ecology.

Andres Guhl is an Assistant Professor at the Development Studies Center (CIDER) at Universidad de los Andes (Bogota, Colombia). He has a B.Sc. in Civil Engineering (Universidad de los Andes), a M.Sc. in Geography (University of Illinois at Urbana-Champaign) and a Ph.D. in Geography (University of Florida). His area of specialty is Latin America, and his research interests include landscape transformations, environment and development, and global change among others. He was an author in the Millennium Ecosystem Assessment (2005) and UNEP's Global Environmental Outlook 4 (2007) reports. He combines GIS and remote sensing with cultural and political ecology to understand landscape change from an integrated perspective. His current research analyzes the relationships between development and landscape change in the Colombian Andes.

Wendy Jepson is an Assistant Professor in the Department of Geography at Texas A&M University. She earned her doctorate from the Department of Geography at the University of California, Los Angeles in 2003. Her research focuses on land-use and land-cover change along South America's modern agricultural frontier, particularly the savannas and dry tropical forests, and the transformation of landscapes in the Lower Rio Grande Valley, Texas. She also examines the relationship between water, society, and equity on the US-Mexico Border.

John Kapito holds a Masters in NGO Management from the University of London, U.K. He co-established the Consumer Association of Malawi in 1994. When the Consumer Protection Law went into force, he became full-time commissioner of the Malawi National Human Rights Commission in Blantyre. His interests include advocacy and lobbying for the social, economic and cultural rights of the voiceless and marginalized.

Eric Keys is an Assistant Professor of Geography at the University of Florida. He earned his doctorate from the Graduate School of Geography at Clark University and worked as a Post-doctoral Researcher at the Center for the Study of Institutions, People, and Environmental Change at Indiana University. He also has degrees from the Institute of Latin American Studies, University of Texas at Austin (M.A.) and Macalester College (Latin American Studies and Spanish). His current research focuses on the land-use and land-cover change in the context of social, economic and environmental change. He has worked in Mexico, Guatemala, the United States and on the United States-Mexico border.

Moussa Koné is a Ph.D. student in the Department of Geography at the University of Illinois Urbana-Champaign. He obtained a postgraduate diploma and masters degree in Geography from the Tropical Geography Institute, University of Abidjan-Cocody, Côte d'Ivoire. His doctoral dissertation research examines the relationships between fire regimes, land-cover change, and greenhouse gas emissions in the Sudanian savanna region of Côte d'Ivoire. This research investigates how farmers and herders use fire as a tool for natural resource management, and how these practices modify landscapes and fire regimes over time. His research is funded by the Geography and Regional Science Program of the National Science Foundation and by the Norman E. Borlaug International Agricultural Science and Technology Fellows Program.

Matthew Koeppe is Senior Project Manager at the Association of American Geographers and Assistant Professorial Lecturer at the George Washington University in Washington D.C. He received his Ph.D. in geography from the University of Kansas in 2005. His research interests include food production, Latin America, Brazil, and environment and development.

Takashi Kurosaki (Ph.D., Stanford University, 1995) is Professor at the Institute of Economic Research, Hitotsubashi University, Tokyo, Japan. Before joining Hitotsubashi University, he worked for the Institute of Developing Economies, Tokyo from 1987 to 1997. His research focuses on microeconomic modeling of low income economies and empirical application of these models to South Asia using household data as well as historical statistics. His recent publications include "The Measurement of Transient Poverty: Theory and Application to Pakistan" (*Journal of Economic Inequality* 4: 325-345, December 2006), "Child Labor and School Enrollment in Rural India: Whose Education Matters?" (*The Developing Economies* 44: 440-464, December 2006), "Consumption Vulnerability to Risk in Rural Pakistan" (*Journal of Development Studies* 42: 70-89, January 2006).

George Malanson is Coleman-Miller Professor in the Department of Geography at the University of Iowa. He earned his doctorate in Geography from UCLA, and has degrees from the University of Utah (M.S. Geography) and Williams College (B.A. Art History). He is a landscape ecologist who has worked on the relations of spatial patterns and processes to plant community composition and diversity using computer simulation methods. He has recently turned attention to deforestation and has served on NSF and NIH panels relevant to land-use change. He is currently Editor for Biogeography for *Physical Geography* and Associate Editor for *Arctic, Antarctic, and Alpine Research*, and he serves on the editorial boards of the *Annals of the AAG*, *Advances in Water Resources*, and *Geography Compass*. He is a Fellow of the AAAS and has received the Parsons Distinguished Career Award in biogeography from the AAG.

Andrew Millington is Professor of Geography and Director of Environmental Programs in Geosciences at Texas A&M University. He was previously Chair of the Dept of Geography at Leicester University, has worked at the Universities of Reading and Sierra Leone, and has been Visiting Professor at University College, Dublin and the University of Gent. He has received B.Sc. (Hull University, 1973), M.A. (University of Colorado, 1977) and D. Phil. (University of Sussex, 1985) degrees. He has researched natural resources issues in western Asia, Africa and South America with funding from the European Union, The World Bank and the UK Natural Environmental Research Council. He was formerly Editor of *The Geographical Journal* and he serves on the editorial boards of the *Annals of the AAG* and *The Geographical Journal.*

Marty Otañez holds a Ph.D. in cultural anthropology from the University of California in Irvine, U.S.A. He is a Post-doctoral Researcher at the Center for Tobacco Control Research and Education of the University of California, San Francisco. His interests include labor, culture, corporate globalization, and the production of film documentaries on tobacco

Chai Podhisita is an Associate Professor at the Institute for Population and Social Research, Mahidol University, Thailand. He obtained his doctorate in anthropology from the University of Hawaii. He is the author of a popular Thai text book on qualitative research methods. His current research interests include peasant society, global householding, youth and

HIV/AIDS. He was awarded the Thailand Research Fund's Citation Award in 2001 and Mahidol University's Best Treatise Award in 2006.

Pramote Prasartkul is Professor of Demography at the Institute of Population and Social Research (IPSR), Mahidol University, Thailand. He is a social demographer interested in population dynamics and the demographic transition in Thailand, including population ageing. He has published descriptions of the age-sex structure of the Thai population, assessments of trends in survival curves, generation life tables, and population projections for the country and is now extending his interest to the study of centenarians in the country. He is also the author of the major demography textbook used in graduate level training in Thailand. He has worked closely with the Carolina Population Center throughout his career, especially in a long-term and still ongoing CPC-IPSR collaborative project based in Nang Rong, in Northeastern Thailand. He received a B.A. in Political Science from Chulalongkorn University and M.A. and Ph.D. degrees in sociology from Cornell University.

Ronald R. Rindfuss is the Robert Paul Ziff Distinguished Professor of Sociology and Fellow of the Carolina Population Center at the University of North Carolina at Chapel Hill, and Senior Fellow, Population and Health Research Program, East-West Center, Honolulu. He has had long-standing research and policy interests in the population and environment area, with an emphasis on land-use change. He also has interests in social demography of fertility and the family, as well as migration patterns in developing countries. His current research interests include migration and environmental issues in Thailand, family change in Japan, and the effect of increased child care availability on fertility in Norway.

Laura C. Schneider is an Assistant Professor in the Department of Geography at Rutgers University. Her research focuses on human-environment relations affecting patterns and processes of land-use and land-cover change. Her specific research interests are monitoring and modeling land transformations, biophysical remote sensing and ecological dynamics of plant invasions in tropical regions.

Stephen J. Walsh is a Professor of Geography, member of the Ecology Curriculum, and Research Fellow of the Carolina Population Center at the University of North Carolina – Chapel Hill. He obtained his doctorate in Geography from Oregon State University in 1977. His current research focuses mapping and modeling land-use/land-cover dynamics, population-environment interactions, biocomplexity in coupled human-natural systems, spatial simulation modeling, spatial digital technologies, and pattern-process relations at the alpine treeline. Major studies are underway in the Ecuadorian Amazon, the Galapagos Islands, northeastern Thailand, and US American West with support from NASA, NSF, NIH, and the USGS. He received the Outstanding Contributions Award and Medal from the Remote Sensing Specialty Group of the Association of American Geographers (AAG) in 1997, Research Honors from the Southeastern Division of the AAG in 1999, National Research Honors for Distinguished Scholarship from the AAG in 2001, and was elected Fellow of the American Association for the Advancement of Science in 2006.

Kenneth R. Young is Professor in the Department of Geography and the Environment, University of Texas at Austin. He is also currently serving as program director of Geography and Regional Science at the National Science Foundation. His research interests are focused on tropical environments, particularly in terms of biodiversity conservation, land-use/land-cover change, and their interrelationships with global environmental concerns. He recently co-edited *The Physical Geography of South America* (Oxford University Press) with Thomas Veblen and Antony Orme.

ACKNOWLEDGEMENTS

In producing an edited book from a conference, editors are acutely aware of the many people who help along the way and we take this opportunity to thank them at this point. First, the organizers of the International Human Dimensions Programme (IHDP) of Global Change Research Community 6th Open Meeting for accepting our proposal to organize a session at their meeting, and secondly all those who presented papers. We also take this opportunity to thank Melinda Paul at Springer for helping us produce this book, and the anonymous reviewers who provided sage advice at the proposal stage. To the authors, thank you for your contributions and for bearing with us as we bombarded you with questions during the reviews and copy-editing of your chapters. Finally, without the unstinting contribution of Indumathi Srinath, one of our cohort of land-change science graduate students in the Geography Department at Texas A&M University, our many errors may not have been spotted, page layouts would have probably have looked awful and the index would never have been produced - thank you Indumathi.

Andrew Millington and Wendy Jepson
College Station, Texas

Chapter 1 – The Changing Countryside

Wendy Jepson and Andrew Millington

Department of Geography, Texas A&M University, MS 3147, College Station, TX, USA 77845

The Changing Countryside

Land use and land-cover change (LULCC) research over the past decade has focused mainly on contemporary primary land-cover conversions in the tropics and sub-tropics, with considerable resources dedicated to the explanation and prediction of tropical deforestation. Over a longer time frame, dating back to the 1970s, the scientific community also focused on the expansion of cultivation and pastoral systems in tropical and sub-tropical drylands, often under the banner of 'desertification.' Generally land change in existing tropical and sub-tropical agricultural regions has been largely ignored. Put succinctly, research projects within the IGBP Land Use Land Cover Change program (which ended in November 2005) centered on transitions from primary vegetation to agriculture or pasture at the expense of similarly detailed attention to dynamics of land change within existing agricultural regions. By ignoring the dynamism in the world's agro-pastoral landscapes, many important drivers and mechanisms of LULCC have been marginalized in recent mainstream land transformation debates.

We organized a session, "Beyond the Primary Transition: Land-Use and Land-Cover Change in Tropical and Sub-tropical Regions," at the International Human Dimensions Programme (IHDP) of Global Change Research Community Open Meeting held in Bonn, Germany, October 2005 to compare contemporary research on land-use and land-cover change in agro-pastoral landscapes to fill this major research gap. This volume is, in part, an outcome of the presentations and scholarly exchanges at the meeting.

We collected and edited several papers from this meeting in this book. We also included a selection of invited contributions from scholars whose work dovetails with our concerns about land change in tropical and subtropical agro-pastoral landscapes.

In addition to addressing an empirical and conceptual gap in land-change research, we selected research that advances some aspect of land-change science. This brand of science seeks to integrate social, biogeophysical, and geographical information sciences to understand the human and environmental dynamics that change the type, magnitude and location of land uses and land covers (Rindfuss et al. 2004; Lambin and Geist 2006). Indeed, the contributions in this volume reflect the three objectives of land-use and land-cover change studies identified in the first phase of the IGBP Land Use Land Cover Change program. Collectively they describe and monitor land-cover changes, explain the processes through which land is altered, and develop spatially explicit models to predict land change. The volume represents how practitioners have integrated, in some combination, the three scientific knowledge realms - social, biophysical, and GIScience - that underpin land-change science. Moreover, they consistently employ analyses that integrate micro-, meso- and macro-scale processes in data and methods, thus confirming the importance and common practice of including multi-scalar approaches in global land-use and land-cover change research.

While these chapters are within the general land-change research paradigm, the contributions may, on an individual basis, diverge and, perhaps, reveal some of the limits or challenges to advancing the narrow scientific agenda outlined by Rindfuss et al. (2004). Some chapters in this volume do not necessarily rely on remotely sensed imagery as data inputs to land-change science or use geospatial analysis to describe and explain land change. Instead, they employ other datasets and forms of statistical analysis appropriate to these other types of data to document and explain land change. For example, persistent cloud cover over the coffee lands of highland Colombia (Chapter 6) limits the use of remotely-sensed data in the reflective spectrum, though not the emitted spectrum, which have been extensively used in land-change science. Instead, a rich agricultural survey provides sufficient data to understand the geographical shifts and changes in coffee production. Similarly, historical studies of land change often preclude spatially explicit GIS analysis linked to individual agents at the parcel- and farm-levels, though they can allow spatially-explicit analysis using various levels of political units. For example, the study of twentieth-century land use in India and Pakistan (Chapter 4), which explains increases in land productivity and crop shifts, depends on applying a statistical decomposition method to a long record of agricultural census data.

The inclusion of these two chapters in this book highlights the potential for employing the 'uncommon' in land-change science such extending the historical view backward using agricultural and other census data, tax records and maps. We do not exemplify the use of remotely sensed data acquired outside the reflective spectrum in the book. Examples of such research exist, but they are far less frequent than those researchers who rely, say, on the Landsat series of sensors. By using the 'uncommon' the research community can extend the spatial and temporal domain boundaries of land-change science (Fig. 1.1).

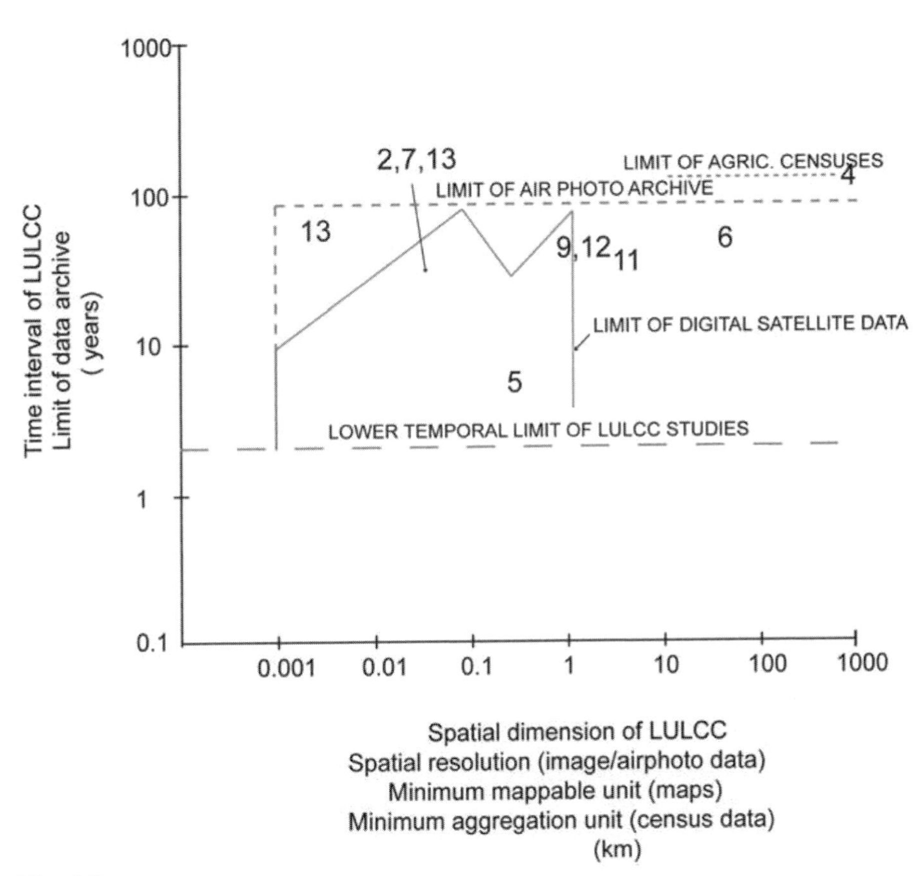

Fig. 1.1

In summary, the work presented in this volume is exemplary of a broader 'portfolio approach' (Young et al. 2006), the use of a toolkit of complementary methods, to study land-use and land-cover change rather than restrictive focus only on the agent-parcel linkage through GIS analysis advocated by Rindfuss et al (2004).

The Contributions

Our contributors are from across the globe and draw from diverse empirical pan-tropical case studies and disciplinary influences. The research reported examines land-use and land-cover change in Bolivia, Brazil, China, Colombia, Côte d'Ivoire, India, Malawi, Mexico, Pakistan, Peru, Senegal and Thailand. Moreover, each chapter advances, either conceptually and/or methodologically, one of three themes present in the collection: (i) adaptations and change in settled agricultural zones, (ii) agricultural intensification, and (iii) markets and institutions. There are five chapters (Chapters 2, 3, 4, 9, and 13) that examine agricultural zones, either long-settled landscapes, such as those in the Andes, or along recent colonization regions in Thailand or Bolivia. A second common theme is agricultural intensification, or the increasing productivity and/or utilization of the land. This results in the modification of land-cover classes rather than transition from one class to another. Four chapters (Chapters 4, 5, 6, and 12) examine these process and the resulting land-use patterns in South Asia (India and Pakistan), Colombia, and Brazil. The third theme addressed in this volume is best described as a discussion of markets and institutions as key drivers of land change in agricultural regions. Four chapters (Chapters 8, 10, 11, and 14) describe land-change outcomes in terms of the interdependent relationship among land tenure, demand from international markets, and new government programs in China, Côte D'Ivoire, Malawi, and Mexico. These chapters demonstrate how markets and institutions, often bound through global commodity chain networks, support and constrain the capacity of farmers in their land-use decisions, leading to diverse land-change outcomes.

Adaptations and change in traditional agricultural areas and recent colonization zones

Global climate change places pressure on small-scale or peasant agricultural systems in the tropical world. ***Kenneth Young*** **(Chapter 2)** examines the question of climate change from the perspective of long-inhabited landscapes in highland Andes. His broad view, which is informed by a conceptual model of long-term dynamism, includes insights into the sustainability, adaptability, and resiliency of these agricultural systems. Young outlines the consequences of climate change whilst paying attention to the technical, conceptual and analytical challenges to the study of global climate on these long-inhabited landscapes. ***Molly Brown*** **(Chapter 3)** also examines climate change on long-inhabited agricultural zones, but narrows the aperture of her empirical investigation, which employs multiple correspondence analyses to quantify the impact of rainfall reductions on income and food security in Senegal. Brown, also informed by concern for sustainability, adaptability, and resilience, demonstrates how climate change induces a shift in resource flows that will affect the viability of local livelihoods. She concludes that while some adaptation is achieved, excessively low rainfall will reduce food security unless there are external interventions that support economic development outside the agricultural economy.

Three chapters examine the biogeographical and spatial dynamism of land-use and land-cover change in colonization regions, areas settled by small-scale farmers in state-sponsored development and settlement projects. The chapters in this volume offer a careful consideration of the consequent, but often overlooked and surprising, land changes that occur after the initial wave of settlement and land conversion. As previous research confirms, household demographics change the labor available for small-scale cultivation that alter land-use patterns. In this volume, we present other economic and biophysical processes that develop or alter alternatives for farmers' production portfolios in these colonization regions.

Human disturbance through land colonization in the southern Yucatán Peninsula not only results in deforestation, but as ***Laura Schneider*** **(Chapter 4)** demonstrates, significant land is modified by subsequent plant invasion. Schneider develops and compares regional and parcel-level spatially explicit models to evaluate the linked socioeconomic and ecological factors that contribute to a bracken (*Pteridium aquilinum* (L.) Kuhn) invasion. Bracken fern invasions impede vegetation succession and shape land availability for cultivation by catalyzing land abandonment which may then lead to further deforestation. She demonstrates that while land modification due to bracken encroachment may cause indirect regional land conversion through deforestation, the best model to predict bracken fern invasion is at the parcel level where richer household data are available. Schneider's chapter confirms that to best predict plant invasions, the analysis needs to go beyond the biophysical dimensions and integrate information from human activities.

***Andrew Bradley and Andrew Millington* (Chapter 13),** challenge the typical tropical lowland colonization model of land change developed in the Brazilian Amazon, through their study of forest change and land-use trajectories in the Chapare of Bolivia. Their analysis, which focuses on coca (*Erythroxylum* coca) cultivation, indicate that the inclusion of illegal activities in a local economy distorts the evolution of households, communities and land-use changes observed in many parts of the Amazon Basin. Using a combination of various types of satellite imagery, household interviews, farm surveys and secondary economic data they have tracked changes in land-use patterns for 45 household/farms and developed a detailed picture of how colonists have responded to what has been at times a buoyant economic market in coca coupled with few restrictions on its cultivation and processing, whilst at other times the area has been subject to strongly enforced anti-narcotics policies.

The process of human occupation evolves over time and alters the spatial imprint of settlement on the landscape. ***Barbara Entwistle* and colleagues (Chapter 9)** take a long view of colonization and land-use and land-cover change in Nang Rong, a colonization zone in Thailand. They draw upon a rich land-use archive (aerial photography) over four decades (1950s-1990s) and many years of experience in the region. The authors elaborate how spatial context (e.g., suitable agricultural land, proximity to rivers, local topography) interact with time-dependent factors (e.g., new roads, international demand for cassava) to influence village settlement and evolution, an accelerating process that shapes key decisions about consequent land use. Indeed, the importance of the international cassava markets parallels the case of coca production in the Bolivian Chapare as a new market beyond the local communities that alters the individual and household land-use decision-making. This chapter, similar to the Schneider's analysis of plant invasions, demonstrates the evolutionary and contingent nature of landscape change after initial settlement. Land changes are path dependent, shaped by marked events such as colonization, but altered over time by the more subtle changes in relations land-users have to social and economic factors.

Agricultural Intensification

Agricultural intensification, the increasing productivity and/or utilization of the land through changes in inputs, labor and/or technology, is the foundation of the world's agricultural revolutions. It has catalyzed the anthropogenic changes in the global environment and provided new avenues for environmental sustainability under increasing population pressures. Intensification during the twentieth century altered ecosystems by excessively loading chemicals, phosphorous and nitrogen into biochemical cycles and, thus, altering biotic interactions in soil and water resources (Matson et al 1997; Bouwman 2005; Tilman et al 2005). Increased cropping intensity, or time a plot of land is under cultivation, also affects regional and global climate dynamics as it changes albedo, ground cover, and carbon sequestration (Keys and McConnell 2005: 325). Yet the same technologies, such as those in the Green Revolution, have also underpinned the tremendous increases in global food production to sustain the growing population, which in turn, according to some, has spared land from further exploitation (Matson 2006).

Two contributions in this volume draw from agricultural surveys to examine national trends in agricultural intensification (India and Pakistan; Colombia), paying particular attention to inter-regional differentiation and crop shifts in land use over meso or long temporal scales. ***Takashi Kurosaki*** **(Chapter 4)** employs a statistical decomposition method to quantify the long-term effects of inter-crop and inter-district crop shifts in productivity for India and Pakistan during the twentieth century. His chapter offers two important perspectives on land-use change in the Indo-Gangetic Plain. Kurosaki concludes that while the aggregate productivity of land increased in the post-independence period for both India and Pakistan, it was not high enough to keep pace with the increasing population resulting in a lower growth rate of agricultural output per capita than per acre.

Andrés Guhl **(Chapter 6)** also uses agricultural survey data to examine the dramatic transformations of coffee intensification (increase use of inputs; reduce plot life; remove shade production) had on the Colombian landscape since 1970. His research confirms the land-sparing outcome: increased yields, which reduced farm areas and created a more agro-diverse and heterogeneous landscape, with more agricultural products and forest regeneration on abandoned lands. Similar to Kurosaki, however, Guhl describes regional shifts of coffee production to places with better agro-ecological conditions. Guhl also clarifies that coffee intensification, which supported middle and larger coffee producers, resulted in loss of shade trees. Therefore, while the overall land in coffee farms has been reduced (land sparing outcomes), the intensive coffee production systems have concentrated in certain regions and lost the ecosystem services provided by the shade tree canopy.

Two other contributions provide analyses on the measurement of increasing productivity and/or utilization of the land. While both chapters are empirically grounded in the Amazonian state of Rondônia, Brazil, they examine agricultural intensification in very different agricultural contexts: capital-intensive soybean production and small-scale agroforesty. ***Wendy***

Jepson, J. Christopher Brown, and Matthew Koeppe **(Chapter 5)** employ satellite remote sensing techniques of moderate resolution data (MODIS) to measure the cropping frequency on agricultural land in soybean-producing areas of Vilhena, a municipality in southern Rondônia. The authors achieve two goals. First, they measure land-use intensification (defined as cropping frequency between 2001 and 2005) from a near biweekly basis, seasonal plant phenological patterns and inter-annual dynamics of croplands to demonstrate that soybeans produced under the most intensive land management regime has increased significantly over five years. Second, the authors demonstrate that the spatial resolution for measuring land modification is sufficient to link to parcel levels. In this case, land tenure (rented or owned) is linked to cropping frequency. Taken together, the research indicates future research opportunities in the expanding areas of mechanized agriculture across tropical South America.

Agricultural intensification in the tropics has value. Agro-forestry, as one example of intensification, is promoted to increase on-farm incomes in hopes of reducing the need to deforest land. ***Keila Aires*** **(Chapter 12)** employs economic methods to analyze the financial viability of two agro-forestry systems. She found that that agro-forestry is profitable and can be recommended as profitable and sustainable land use, confirming the land-saving claims promoted by agricultural development.

Markets and Institutions

Synergies between markets and institutions lead to dramatic and contested changes in the land across the tropics. Markets and institutions are intimately linked through the production and movement of commodities. They provide opportunities or restrict farmers in their production with incentives or disincentives to modify or completely change their land-use practices. Four chapters explore this interdependence among domestic and international markets, formal and informal institutions, and farmers.

Land tenure frames land-use decisions. ***Thomas Bassett and Moussa Koné*** **(Chapter 8)**, examine how changes in land law, within the context of political-economic crises, have led to the expansion of cotton and a diversification of crops from row (cotton) to tree crops (cashews and mangos). The authors highlight how moves toward land privatization, supported by the World Bank, have led to increased tree planting as a means to establish land rights and, thus, access private land title. This concern about land access, control and use on the brink of privatization has also reconfigured the relations between pastoral and farming communities, where land rents have shifted from in-kind or symbolic gifts to cash payments,

such as grazing fees. These synergies between markets (cotton economy) and institutions (land law) have led to changes in rural relations and land-change outcomes as farmer strategies to cope with this agrarian change.

The link between markets and institutions is less obvious, but nonetheless, powerful in ***Laura Ediger*'s (Chapter 11)** analysis of afforestation projects in western Yunnan province, China, where the upper watershed boundaries of the Mekong, Nujiang, and Yangtze Rivers are located. Land decisions in China have been devolving from state to household as a move to increase land-use efficiency and production in support of the country's overall economic growth. The state has attempted to decentralize forest resource management; the household and existing collective management mechanisms have been employed to support pine plantation development. Ediger compares the incentive structure of both systems and describes the real land-change outcomes of each. She concludes that the more flexible and local collective arrangements, rather than the household-bases management system, distribute risk and costs. Collective arrangements reduce the pressure on the household economic system and realigns the farmers' incentives to the state's goal of increasing environmental services.

***Eric Keys* (Chapter 10)**, drawing from extensive interviews and fieldwork, provides a sharp analysis of how commodity chain and domestic market relations for jalapeño peppers (chili) drive greater land-cover conversion than traditional subsistence farming on the Yucatán Peninsula, Calakmul, Mexico. Keys outlines complex land-change dynamics in Calakmul. The rapid adoption of chili cultivation, both mechanized and swidden-base production, among farmers in the frontier region is the only significant commodity for income generation. The reliance on chilis, with its price vulnerabilities and distribution structure (oligopsony), leads to two inter-related land changes. First, profits from chili production can be invested in more production to develop the farmers' land-use portfolio. Moreover, participation in the chili market encourages farmers in the region to expand the cultivation of their commodity and subsistence crops to hedge against price fluctuations. Keys deepens his analysis of farm-to-market relationships, describing how the middlemen (*coyotes*), support or constrain the capacity of a farmer to diversify his chili production, and thus influencing farmers' decisions on land use.

In the final chapter, ***Helmut Geist, Marty Otañez and John Kapito* (Chapter 14)** target tobacco in Malawi to trace how a globalized, yet characteristically neocolonial, commodity chains and institutions have transformed local environments, causing land degradation in existing agricutlrual areas and new clearance of the dry forest called *miombo*. The authors illustrate how global tobacco markets and the institutions that develop from the tobacco economy, such as marketing boards, trade organizations and development agencies, provide incentives to produce the crop regardless of other negative consequences on forest resources, economic development, and human health.

References

Bouwman AF, Van Drecht G, Van der Hoek KW (2005) Global and regional surface nitrogen balances in intensive agricultural production systems for the period 1970-2030. Pedosphere, 15(2):137-155

Keys E, McConnell WJ (2005) Global change and the intensification of agriculture in the tropics. Global Environmental Change-Human and Policy Dimensions, 15 :320-337

Lambin EF, Geist HJ (Eds.) (2006) Land-Use and Land-Cover Change: Local Processes and Global Impacts. Springer, Boston

Matson PA, Parton WJ, Power AG, Swift MJ (1997).Agricultural intensification and ecosystem properties. Science 277(5325): 504-509

Matson PA, Vitousek PM.(2006).Agricultural intensification:Will land spared from farming be land spared for nature? Conservation Biology 20(3): 709-710

Rindfuss RR Walsh SJ, Turner BL, Fox J, Mishra V (2004).Developing a science of land change:Challenges and methodological issues. In: Proceedings of the National Academy of Sciences of the United States of America 101(39) pp 13976-13981.

Young OR, Lambin EF, Alcock F, Haberl H, Karlsson SI, McConnell WJ, et al. (2006). A portfolio approach to analyzing complex human-environment interactions: Institutions and land change. Ecology and Society 11(2):31 [online http://www.ecologyandsociety.org/vol11/iss2/art31/]

CHAPTER 2 - Stasis and Flux in Long-Inhabited Locales: Change in Rural Andean Landscapes

Kenneth R. Young

Department of Geography & the Environment, University of Texas at Austin, TX 78712, USA

ABSTRACT

Stasis is as interesting as flux in anciently settled landscapes, especially given future shifts driven in part by global climate change. The goals of an expanded research agenda could involve asking how much history can explain current locations and types of stability versus change, if land-use shifts necessarily result in altered landscape patterns, whether multiple stable states are possible, and if landscape responses lag in predictable ways in regards to social and biophysical drivers. For example, the rural landscapes of the tropical Andes Mountains were settled several millennia ago. The cumulative effects of millennial land use have altered soil attributes, eliminated certain plant and animal species, and removed much of the original land-cover types, implying that patterns and processes will be different than in sites undergoing recent colonization and land conversion. Detecting and explaining change in these kinds of landscapes will be demanding tasks given the fine grain of landscape patches, the prevalence of afforestation with non-native trees, and the relatively rapid responsiveness of local land use systems. Climate-driven forcing acting upon rural livelihoods will need to be monitored and mitigated. Long-inhabited landscapes represent a distinct challenge for the study of change, but also have much to teach in terms of long-term sustainability and adaptability.

INTRODUCTION

Biogeographical and socio-economic shifts resulting from global climate change and local land-use feedbacks are already altering agricultural and pastoral systems in the tropical Andes Mountains (Table 2.1). The most visible change is in the disappearance of small glaciers and the retreat of the lower margins of ice caps (Thompson et al. 2003, 2006)[1], associated with negative glacial balance, warmer temperatures, and in some places less precipitation (Kaser and Osmaston 2002; Vuille et al. 2003; Mark and Seltzer 2005). However, there are many additional implications (Carey 2005; Bradley et al. 2006; Pounds et al. 2006; Young and Lipton 2006), some of which will be explored further in this chapter.

Contemporary changes in many utilized landscapes around the world have begun to show at least subtle responses to the global environmental forcing caused by increased greenhouse gases (Houghton et al. 2001)[2]. The most profound alterations may be expected in coastal areas or islands near sea level (Barnett and Adger 2003; Ericson et al. 2006), and in high latitude and high altitude sites (Barry 2006; Peterson et al. 2006). Simultaneously, many tropical and subtropical landscapes are being converted by colonization and commercial agriculture and, as a consequence, are being reorganized by increased economic connections (Bawa and Dayanandan 1997; Geist and Lambin 2002; Etter et al. 2006). This chapter, however, examines a relatively overlooked landscape category in this context. This type consists of locales where human occupation is ancient, that is, where colonization and land-cover transformation occurred centuries to millennia ago. The case evaluated here is for the tropical Andes, but similar situations may be found in the Mediterranean, the Near East, and in the tropical Asian and Mesoamerican highlands (Whitmore and Turner 2001; Fedick and Morrison 2004; Butzer 2005). These are places where agriculture was already developing by the mid-Holocene

---

[1] http://www-bprc.mps.ohio-state.edu/Icecore/front-page.html

[2] http://www.ipcc.ch

Table 1. Climate-driven changes and consequences reported for the tropical Andes.

Biophysical Change	Consequence	Reference
Increased CO_2	Initial increase in net primary productivity; favors some plant species	Young and Lipton 2006
Warmer temperatures	Shifts ecological zones upward	Vuille et al. 2003
Altered precipitation	Increases glacial retreat	Kaser and Osmaston 2002
Less hail and sleet	Local perceptions of less damage to crops	Young and Lipton 2006
Altered seasonality	More area burns in dry season	Young and Lipton 2006
Altered hydrology	Less available water	Young and Lipton 2006
Less ice cover	Open non-glaciated habitat appears	Silverio and Jaquet 2005; Ramírez et al. 2001
Negative glacier mass balance	Lakes and wetlands grow from meltwater; increased risk from unstable ice and dammed lakes; less environmental buffering; restricts irrigated agriculture	Postigo, 2006; Kaser and Osmaston 2002; Vilimek et al. 2005; Young and Lipton 2006
Increased rate of glacial retreat	Increased rate of environmental change	Thompson et al. 2003
More exposed substrate	Allows primary succession to begin; vegetation appears in satellite images in places previously ice covered	Postigo 2006
Higher cloud bank	Elevation is higher at which mist forms in cloud forests	Pounds et al. 2006
Altered forest cover	In some place more woody plants	Kintz et al. 2006
Altered grazing areas	Different lands used for grazing; need to redistribute grazing lands	Young and Lipton 2006; Postigo 2006

It is likely that there are lessons to be learned from these long-inhabited agricultural landscapes, including insights and long-term perspectives on sustainability, resilience, and adaptability. The tropical Andes are particularly remarkable in that they combine globally outstanding values for native biological diversity, including many endemic plant and animal species (Myers et al. 2000; Brooks et al. 2002), with the maintenance of livelihoods of numerous indigenous peoples (Maybury-Lewis 2002; Radcliffe and Laurie 2006). This chapter begins by developing a conceptual model of the dynamism of spatial aspects of rural Andean landscapes, and then examines some of the implications for the types of change that occur and that do not occur. Next, the consequences for climate change are evaluated in relation to this model. Finally, linkages and extrapolations are made to other ancient landscapes.

RURAL ANDEAN LANDSCAPES

Andean farmers and pastoralists are managers of resources and risks. They respond to a high mountain environment in a variety of ways, including specialization on particular sets of crops and domesticated animals if their lands are found at one elevation or in one particular climatic regime. More typically, households and communities occupy and cultivate lands scattered along an altitudinal and/or topographic gradients, which act to provide access to a variety of agroecological regimes, hence diversifying the resource base available (Troll 1968; Brush 1976; Denevan 2001). Zimmerer (1999) reviewed and updated a strict zonal approach to Andean mountain agriculture by describing and labeling actual patterns as "overlapping patchworks." Agricultural fields are not arrayed neatly up and down the slopes, and elevation is not the only, or even the primary environmental control, on crop choice in many areas (Figure 2.1). In fact, farmers maintain 3 to 12 or more fields, and even then, they may multi-crop and often follow fallow periods by planting yet a different crop. Many households also have a variety of domesticated animals, including livestock that are grazed on fallowed lands and in high elevation grasslands (Figure 2.2).

One way to implement this landscape mosaic model is through the use of the pattern metrics of landscape ecology (McGarigal and Marks 1995; Neel et al. 2004; Turner 2005), which quantify the numbers, relative importance, shapes, and spatial patterns of the landscape patches. Kintz et al. (2006) recently did this for 11 different land use/land cover (LULC) classes that correspond to the vegetation types (several kinds of forests and of shrublands, rocklands, grasslands, wetlands) and agricultural areas in a 30 x 40 km northern Peruvian site previously described by Young (1993). Their data came from Thematic Mapper (TM, ETM+) satellite images from 1987 and 2001, so change over 14 years could also be quantified. They found a total of 28,053 LULC patches in 1987, which reduced to 15,659 patches in 2001 through human-caused processes such as the cutting of forests and the establishment of new agricultural fields, and through ecological processes such as the invasion of grasslands by shrubs. These changes coincided with, and were probably driven by, in-migration and population growth during that time period, which more than doubled the number of local inhabitants.

Fig. 2.1. Landscape mosaic of crops, planted trees, and natural environmental heterogeneity at about 3400 m elevation in Ancash, north-central Peru

Neither the spatial resolution (30 m) nor the spectral discrimination (seven bands) of these satellite-borne sensors permitted landscape analyses that fit the fine-grain approach that farmers usually take to their lands in the Andes. They may, for example, plant dozens of different potato varieties within the same field (Zimmerer 1996), potentially allowing for microsite selection that would take into account within-field edaphic heterogeneity. Future development of remote sensing and GIScience approaches (Brunn et al. 2004; Young and Aspinall 2006) undoubtedly will provide digital and mapped data on land cover that is on the same scale as are the perceptions, tenures, and uses of the local people. A specific example of the kind of research needed to match sensor system capabilities with detectable and quantifiable ecological and land use/land cover change (LULCC) processes is the recent evaluation of biomass burning in highland Peru and Bolivia done by Bradley and Millington (2006). This evolving GIScience and landscape ecology approach to Andean locales can also be enriched by comparison with such approaches in other kinds of landscapes (Walker 2003), and by using methods that incorporate historical legacies (Lunt and Spooner 2005).

As examples of such legacies in the Andes there are often rock walls and hedges in the landscapes that delimit land demarcations (Figure 2.1). If ownership or usership rights have changed through time, then these features may no longer have a function, and instead become part of otherwise uninterpretable landscape heterogeneity. It is common to find terracing, irrigation canals, drained fields, causeways, and even hilltop archaeological sites that are not part of current land use. In areas subjected to surface surveys in the Andes, such features are often found to date back as far as several thousand years (e.g., Stanish 1989; Miller and Gill 1990; Bermann and Castillo 1995; Lau 2002, 2005). It is likely that some soil features, at least depth and slope angle, described for the mollisols found on terraced slopes in the Peruvian Andes (Eash and Sandor 1995; Sandor and Eash 1995; Inbar and Llerena 2000), are also legacies from the past, due to slope modifications and soil amendments made 500 to 1000 years previously. Foster et al. (2003) recently called on ecologists and conservation biologists to pay more attention to how past land-use legacies can continue to alter current conditions, taking examples from Puerto Rico and the northeastern United States. This topic has already received much attention from geographers (Gade 1999; Doolittle 2000; Denevan 2001).

Farmers are also affected in decision making by their needs or wishes for products that they cannot make for themselves. In the Andes, there is a gradient from households with independent subsistence to farmers completely dedicated to cash crops. Improvements of transportation infrastructure and the rapid growth of urban populations create many new opportunities for selling produce (Rudel and Richards 1990; Chant 1998; Wiegers et al. 1999; Bebbington 2000). There is also partial out-migration wherein some family members move to cities for employment or education opportunities, but send remittances back to the countryside, perhaps also combined with the facilitation of the sale of agricultural products in the city. These economic and personal ties are reminders that the monetary and information flows that characterize globalization (e.g., Salisbury and Barnett 1999; Mahutga 2006; Zimmerer 2006) are also felt in the smallest settlements of the Andes.

Fig. 2.2. Decisions on farming and grazing are often made based on household conditions and in reference to community dynamics. Shown is a house and associated lands at 4900m elevation in southern Peru.

STASIS VERSUS FLUX

Land-Change Science (Gutman et al. 2004) and studies of LULCC (Lambin et al. 2003) contain the word *change* in their raison d'être. If flux of a variable or in a mapped element is detected using remote sensing or other GIScience techniques, then a variety of methodological and theoretical approaches are used to try to elucidate the biophysical, socio-economic, or coupled processes that may explain the change(s). Proximate causes are sometimes distinguished from ultimate ones to help clarify complex and indirect causality (Lambin et al. 2001).

So, change in land cover and landscape pattern indicates a process or processes at work (Aplin 2004), while the presence of expected processes can be evaluated by looking for predictable changes in landscape patterns and metrics through time (Crews-Meyer 2002). As Kintz et al. (2006) pointed out for the Andes, and many others have mentioned for other landscape studies in mountainous terrain (Colby and Keating 1998; Bishop et al. 2003; Quincey et al. 2005; Brandt and Townsend 2006; Wulder et al. 2006), sometimes important processes cannot be detected on the surface of the Earth or by using particular sensor systems. Or other difficulties, such as shadows, clouds, or unavailable imagery from many past time periods, may limit change detection studies.

As part of the dynamism, there will be important seasonal fluxes in Andean landscapes as rainy and dry seasons pass by; as farmers respond by plowing, planting, and harvesting; and as pastoralists practice transhumance by moving their animals to higher lands during the dry seasons and to lower elevations during the rains. There also are many cyclical controls on landscape change that are subdecadal in duration, including those driven by ENSO (El Niño Southern Oscillation; Vuille et al. 2000a, 2000b; Caviedes 2001) and also many crop patterns that change through two to eight-year fallow cycles. Separating these kinds of short term fluxes from the directional or progressive changes imparted by longer term land-cover transformations associated with land use and with the changes forced by climate change is an important research concern. In addition, the vegetation-change processes that cause natural disturbances and result in ecological succession can cause multiple successional trajectories to develop on a given landscape, with recently disturbed vegetation adjacent to pockets of recovering and undisturbed remnant vegetation. In most inhabited Andean landscapes below 3500 m, the landscape matrix consists of shrublands, while at higher elevations the matrix often consists of grasslands or mixed tropical alpine vegetation (Young et al. 2007). Some reciprocal transitions can potentially be detected as fallowed fields are allowed to fill with woody plants while nearby Andean forests may be cut to allow for cultivation.

Nevertheless, there are also times when stability, such as cases or places of "no change" in change detection studies, can be as interesting, as informative, and as important for landscape analyses as these examples of flux. Stable parts of the Andean landscape studied by Kintz et al. (2006; see also Kintz 2003) made up 63 percent of the surface area and included such important elements as core patches of Andean forest that were not affected by deforestation and also tracts of land kept under constant cultivation. For Brandt and Townsend (2006), working with Thematic Mapper imagery from southeast Bolivia, the values of land-cover change from 1985 to 2003 were 17 percent at around 1400 m to 3000 m elevation, and 7 percent at higher elevations up to 4600 m. In other words, 83 to 93 percent of the landscapes were considered stable using their methods.

Stable landscape elements represent sites for local people that constitute forest reserves available to harvest firewood, find medical herbs, hunt white-tailed deer, or serve as places available in order to expand agriculture at some time in the future. Alternatively, they may represent areas that are used for intensive agriculture, maintained with fertilizers or with frequent crop rotations. Orchards and home gardens are long-lasting features on the landscape. Communal grazing areas might also be relatively stable through time, at least in terms of their outer limits. The resulting spatio-temporal heterogeneity provides opportunities to maintain some native biodiversity on the landscape, including predators such as spiders and insectivorous birds that may be beneficial for controlling agricultural pests.

It is also possible to detect stability in a LULCC study that is not due to an absence of a driver of change, but rather to the consequence of having two or more opposing forces that cancel each other out. For example, a landscape affected by equal amounts of deforestation and afforestation may show no net forest loss, despite important change processes at work. It is desirable for change detection studies to include mapping out of the different change and stability classes to check for these kinds of artifacts and for other complications that may affect interpretation.

These mixed signals of flux and stasis complicate the landscape patterns that need to be examined. But they also come closer to capturing the true imprint of human land use upon agricultural landscapes. Thus, there is probably no substitute for doing LULCC studies for as many time periods as possible, for extracting data on both change and stability (Crews-Meyer 2004), for including historical legacies by mapping in pre-existing but unutilized cultural landscape features, and by paying special attention to the places and from-to classes that suggest stability. Conceptually, stasis might represent what is most predictable to Andean farmers, while flux may signal either transitional opportunities or possibly losses due to disturbance or due to change towards terminal land-cover states, for example, to exposed rock or to urban uses.

CONSEQUENCES OF CLIMATE CHANGE

Climate is always changing, sometimes erratically around long-term means, but in times of global environmental change with directional shifts in temperature, precipitation, and resulting soil moisture regimes. Recent discoveries by Thompson et al. (2006) in the form of 5100 year old plant remains exposed by the retreat of the Quelccaya Ice Cap in southern Peru show that current warm conditions in the central Andes are novel compared to the previous five millennia. Close by, Bush et al. (2004) extracted a paleoclimate record of almost 50,000 years in age from lake sediments that showed surprisingly little change. The changes that have occurred in land cover in that part of South America due to recent human-caused deforestation and burning (Archard et al. 2002) dwarf the magnitude, intensity, and speed of the vegetation shifts implied by their long-term paleoclimatic record.

Natural selection acts upon the evolution of plant and animal species over millennial time scales. Thus, contemporary and rapid climate change may potentially put some Andean species at risk of extinction if their physiological limits are exceeded and if limitations on dispersal prevent their establishment in alternative sites. Widespread land-cover alterations by people began long ago in the tropical Andes (Ellenberg 1979; Young 1998; Gade 1999) and would have had the effect of exacerbating the isolation and fragmentation of distributions of some kinds of species. For example, Bush et al. (2005) found evidence in fossil pollen proxies of widespread anthropogenic landscape alteration in northern Peru beginning 3500 years ago. This means that species sensitive to burning and other habitat modifications would have been altered in their distributions and abundances, while other species might have been favored. Even earlier in the history of human occupation of the Andes, many large mammals went extinct, coincident with climate change in the early Holocene and with the arrival of hunters and gatherers (Stahl, 1996). Current climate change is thus acting upon a subset of the original Andean flora and fauna, in many cases with species relatively robust or resilient in the face of the human-caused landscape modifications, at least as judged by their continued persistence. Exceptions will necessarily be conservation priorities and may require active intervention and management (Young 1997; Young et al. 2002).

On a landscape scale, the climatically-induced and human-caused shifts in distributions of plants and animals will, in turn, cause changes in the vegetation and ecosystem types that they constitute. Some vegetation types, for example, will contract, become perforated, or otherwise be reduced in dominance. Others made up of species pre-adapted to survive under grazed, burned, or fallowed conditions will increase in cover. In fact, most Andean landscapes below 3500 m elevation are currently dominated by shrublands, with multiple-stemmed woody plants that can resprout following disturbance or under unfavorable conditions. Above 3500 m, basal-growing cespitose graminoids predominate, with sedges in wetlands and grasses on most well-drained substrates. The directional change in current climate change should be detectable as spatial shifts in ecological transition zones (ecotones) (Young and León 2007). In effect, Kintz et al. (2006) report increased shrub invasion of high elevation grasslands in northern Peru, which might be an early signal of future climate-caused shifts to be expected in the tropical Andes.

Wavelet analysis of magnetic susceptibility was used by Bush et al. (2005) to statistically distinguish repeated 210-year climate cycles of increasing and decreasing precipitation from other indicators of change in the last 8000 years of the record provided by the Andean lake sediments they studied. This kind of study offers many opportunities for further clarifying the relative roles of climate change (in this case presumably caused by a solar cycle) and of land use in affecting Andean landscapes over century-scale time intervals. Study of plant remains and fossils in cultural or other dated contexts (Seltzer and Hastorf 1990; Chepstow-Lusty and Winfield 2000) provide independent lines of evidence on human livelihoods, with implications for interpreting Andean landscapes. Perry et al. (2005) proffer a recent example with their documentation of the use of an Amazon-derived cultigen (Marantaceae) in farming at an Andean site on the Pacific drainages of southern Peru about 4000 years ago.

One conspicuous current feature of rural Andean land use, and one that is likely also to have been characteristic during past millennia, is the capacity for people to make subtle spatial and temporal shifts in relation to climate and seasonal signals. Farmers can choose from dozens or more varieties of major crop species, can plant or fallow their different fields based on expected harvests, and can supplement their subsistence with purchases or from products derived from poultry or livestock. They thus can manage risks and expectations, using the redundancy of their land use practices to provide margins and flexibilities. Cues of future shifts may be taken from the appearances of clouds, winds, certain plants and animals, or other events; some of these may be folklore, some certainly do contain information useful for making short-term climate predictions (e.g., accurate ENSO predictions made by local people in southern Peru through stargazing, Orlove et al. 2000). Weather prediction more than several months ahead may not be possible, but Andean farmers also have access to collective memories and traditions, which may point to given fields or crops being most likely to produce well under certain conditions based on past experience.

Given these capabilities, it is not hard to imagine that land use may change rapidly, on the order of within 2 to 12 week time intervals. For example, pastoralists noting that plants are now growing on recently exposed glacial drift, may move their herds of sheep and llamas to take advantage of the new foraging opportunities (Figure 2.3). Farmers may decide to irrigate their fields for cash crops if increased water is available that year due to glacial melt. Alternatively, especially in Andean sites below newly deglaciated mountaintops, farmers may switch to rain-fed agricultural methods. All of these decisions may be constantly revisited and practices altered given the length of the dry season or rainy seasons, if pests such as locusts appear, if diseases strike (Salazar 2006), or if prices in urban markets send other signals to the producers.

Information to be used in land-use decision making is drawn from traditions, from personal experiences, from outside sources such as NGOs (non-governmental organizations), mass media, and field observations. Often communal decisions are made through consensus, leavening the needs of families with the rights acquired through communal labor or through other contributions. Increased communication due to easier travel and to the availability of more telephone and internet connections, potentially allows the sharing of experiences and opinions among Andean communities. Sometimes these interconnections are fostered by NGOs (Bebbington and Kothari 2006). Simultaneously, however, other cultural and socio-economic shifts are fomenting out-migration of young people to urban areas. Capacity to respond to likely future agro-pastoral alterations caused by climate change is associated with all these factors. Much decision making is done at a household level, but obviously is also influenced by community dynamics and by external factors such as degree of isolation from market forces (Mayer 2002; Young and Lipton 2006).

Stability of some biophysical and social elements of land use may also be important. Successful passing of knowledge about farming to children acts to maintain practices that have proven useful for land use in the past (Young 2002). In turn, stable patches in the shifting landscape mosaics of the Andes may serve as refugia for vegetative regrowth or for wild plants and animals that otherwise would not survive in an intensively farmed landscape. In this case, the redundancy is ecological and may not be part of deliberate or planned land use practices. The result, however, is similar, providing core areas of habitat that are available for future use or as sources for ecological succession.

Fig. 2.3. Pastoralists decide where and when to graze, and often manage herds of several species with different foraging needs, including sheep, alpacas, and llamas. Shown is a mixed flock with its owner who uses lands above 4000 m elevation in Huancavelica, central Peru.

CHANGE IN ANCIENT LANDSCAPES

Because of the antiquity and pervasiveness of human influence, long-inhabited landscapes undoubtedly share some biological, edaphic, geographical, and archaeological features. The ways that change occurs and the consequences thereof will be very different from change expected on new agricultural frontiers. As shown by examples in this chapter, useful information can come from historical approaches, in addition to the more typical combination of ecological and socio-economic methods used in the study of LULCC (Walsh and Crews-Meyer 2002).

As a group, these landscapes were the scenes of origins of human history and the sources of animal breeds and of plant crop varieties. So they have current merit and interest for agricultural conservation efforts, although that may require both *ex situ* and *in situ* approaches (Brush, 2000). In addition, there are interesting possibilities for ecological restoration (for the Andes, see Sarmiento 1997, 2000; Rhoades et al. 1998; Young 1998; Groenendijk et al. 2005) using remnant vegetation for inspiration and as source materials for reforestation.

Interpreting these landscapes requires the evaluation of continuing influences of past events, including species extinctions, subsequent biotic reorganization, soil erosion, and ancient deforestation and desertification. Past events still affect current processes, creating land use and land cover legacies that continue to produce both spatial and temporal contingencies and both stasis and flux.

IMPLICATIONS AND CONCLUSIONS

There have been centuries of unfair resource extraction in the high Andes that have diverted natural and human resources down to urban settlements at lower elevations or to foreign lands. In addition, the places most affected by ongoing climate change in the tropical latitudes are the highlands. There are thus practical and ethical reasons to be concerned about Andean agricultural landscapes, in addition to the search for general lessons to be learned from these ancient landscapes in relation to contemporary agricultural landscapes. Some of the landscape transformations of concern in tropical lowlands today, for example, already occurred 3000 to 8000 years ago in the tropical highlands and in the Mediterranean. The primary transition passed through long ago.

In terms of needed research, it is often hard to measure LULCC in the Andes Mountains due to technical, conceptual, and analytic difficulties. Additional sensor resolution and computational dexterity would help. By comparison with other ancient landscapes, it might be possible to better understand feedbacks, positive and negative. The biogeographical consequences of contemporary change in countrysides of the tropics have been outlined by Daily and collaborators working in the middle elevations of Costa Rica, and in particular on relations to the diversity of native species of mammals (Daily et al. 2003), birds (Daily et al. 2001; Hughes et al. 2002), invertebrates (Ricketts et al. 2001; Goehring et al. 2002; Horner-Devine et al. 2003), and plants (Mayfield and Daily 2005; Mayfield et al. 2006). This chapter points to many additional concerns raised by the inclusion of Andean case studies and by looking more directly at the land use/land cover consequences involved. Climate change is only one of several causes of transformations occurring in Andean landscapes, although the others are mostly socio-economic in nature. As a result, it is also important to consider effects of markets opening, of demographic shifts, and of globalized information flows.

REFERENCES

Aplin P (2004) Remote sensing: Land cover. Progress in Physical Geography 28:283-293

Achard F, Eva HD, Stibig HJ, Mayaux P, Gallego J, Richards T, Malingreau JP (2002) Determination of deforestation rates of the world's humid tropical forests. Science 297:999-1002

Barnett J, Adger WN (2003) Climate dangers and atoll countries. Climatic Change 61: 321-337

Barry RG (2006) The status of research on glaciers and global glacier recession: A review. Progress in Physical Geography 30:285-306

Bawa KS, and Dayanandan S (1997) Socio-economic factors and tropical deforestation. Nature 386:562-563

Bebbington A (2000). Reencountering development: Livelihood transitions and place transformations in the Andes. Annals of the Association of American Geographers 90:495-520

Bebbington A, Kothari U (2006) Transnational development networks. Environment and Planning A 38: 849-866

Bermann M, Castillo JE (1995) Domestic artifact assemblages and ritual activities in the Bolivian Formative. Journal of Field Archaeology 22:389-398

Bishop MP, Shroder Jr. JF, Colby JD (2003) Remote sensing and geomorphometry for studying relief production in high mountains. Geomorphology 55:345-361

Bradley AV, Millington AC (2006) Spatial and temporal scale issues in determining biomass burning regimes in Bolivia and Peru. International Journal of Remote Sensing 27:2221-2253

Bradley RS, Vuille M, Diaz HF, Vergara W (2006) Climate change: Threats to water supplies in the tropical Andes. Science 312:1755-1756

Brandt JS, Townsend PA (2006) Land use-land cover conversion, regeneration and degradation in the high elevation Bolivian Andes. Landscape Ecology 21:607-623

Brooks TM, Mittermeier RA, Mittermeier CG, da Fonseca GAB, Rylands AB, Konstant WR, Flick P, Pilgrim J, Oldfield S, Magin G, Hilton-Taylor C (2002). Habitat loss and extinction in the hotspots of biodiversity. Conservation Biology 16: 909-923

Brunn, S.K., S.L. Cutter, and J.W. Harrington, Jr. 2004. Geography and Technology. Kluwer Academic, Dordrecht.

Brush SB (1976) Man's use of an Andean ecosystem. Human Ecology 4:147-166

---------- (2000) Genes in the Field: On-Farm Conservation of Crop Diversity. Lewis Publishers, Boca Raton, Florida

Bush MB, Hansen MBBCS, Rodbell DT, Seltzer GO, Young KR, León B, Abbott MB, Silman MR, Gosling WD (2005) A 17,000 year history of Andean climate and vegetation change from Laguna de Chochos, Peru. Journal of Quaternary Science 20:703-714

Bush MB, Silman MR, Urrego DH (2004) 48,000 years of climate and forest change in a biodiversity hot spot. Science 303: 827-829

Butzer KW (2005) Environmental history in the Mediterranean world: Cross-disciplinary investigation of cause-and-effect for degradation and soil erosion. Journal of Archaeological Science 32:1773-1800

Carey M (2005) Living and dying with glaciers: People's historical vulnerability to avalanches and outburst floods in Peru. Global and Planetary Change 47:122-134

Caviedes CN (2001) El Niño in History: Storming through the Ages. University Press of Florida, Gainesville

Chant S (1998) Households, gender and rural-urban migration: Reflections on linkages and considerations for policy. Environment and Urbanization 10:5-21

Chepstow-Lusty A, Winfield M (2000) Inca agroforestry: Lessons from the past. Ambio 29: 322-328

Colby JD, Keating PL (1998) Land-cover classification using Landsat TM imagery in the tropical highlands: The influence of anisotropic reflectance. International Journal of Remote Sensing 19:1479-1500

Crews-Meyer KA (2002) Characterizing landscape dynamism using paneled-pattern metrics. Photogrammetric Engineering and Remote Sensing 68:1031-1040

---------- (2004) Agricultural landscape change and stability in northeast Thailand: Historical patch-level analysis. Agriculture, Ecosystems and Environment 101:155-169

Daily GC, Ceballos G, Pacheco J, Suzán G, Sánchez-Azofeifa GA (2003) Countryside biogeography of neotropical mammals: Conservation opportunities in agricultural landscapes of Costa Rica. Conservation Biology 17:1814-1826

Daily GC, Ehrlich PR, Sánchez-Azofeifa GA (2001) Countryside biogeography: Use of human-dominated habitats by the avifauna of southern Costa Rica. Ecological Applications 11:1-13

Denevan WM (2001) Cultivated Landscapes of Native Amazonia and the Andes. Oxford University Press, New York

Doolittle WE (2000) Cultivated Landscapes of Native North America. Oxford University Press, Oxford

Eash NS, Sandor JA (1995) Soil chronosequence and geomorphology in a semi-arid valley in the Andes of southern Peru. Geoderma 65: 59-79

Ellenberg H (1979) Man's influence on tropical mountain ecosystems in South America. Journal of Ecology 67:401-416

Ericson JP, Vörösmarty CJ, Dingman SL, Ward LG, Meybeck M (2006) Effective sea-level rise and deltas: Causes of change and human dimension implications. Global and Planetary Change 50:63-82

Etter A, McAlpine C, Phinn S, Pullar D, Possingham H (2006) Characterizing a tropical deforestation wave: A dynamic spatial analysis of a deforestation hotspot in the Colombian Amazon. Global Change Biology 12:1409-1420

Fedick SL, Morrison BA (2004) Ancient use and manipulation of landscape in the Yalahau region of the northern Maya lowlands. Agriculture and Human Values 21: 207-219

Foster D, Swanson F, Aber J, Burke I, Brokaw N, Tilman D, Knapp A (2003) The importance of land-use legacies to ecology and conservation. Bioscience 53:77–89

Gade DW (1999) Nature and Culture in the Andes. University of Wisconsin Press, Madison

Geist HJ, Lambin EF (2002) Proximate causes and underlying driving forces of tropical deforestation. Bioscience 52: 143-150

Goehring DM, Daily GC, Şekerçioğlu CH (2002) Distribution of ground-dwelling arthropods in tropical countryside habitats. Journal of Insect Conservation 6:83-91

Groenendijk JP, Duivenvoorden JF, Rietman N, Cleef AM (2005) Successional position of dry Andean dwarf forest species as a basis for restoration trials. Plant Ecology 181:243-253

Gutman G, Janetos A, Justice C, Moran E, Mustard J, Rindfuss R, Skole D, Turner II BL (2004) Land-Change Science: Observing, Monitoring, and Understanding Trajectories of Change on the Earth's surface. Kluwer, New York.

Horner-Devine MC, Daily GC, Ehrlich PR, Boggs CL (2003) Countryside biogeography of tropical butterflies. Conservation Biology 17:168-177

Houghton JT, Ding Y, Grigs DJ, Noguer M, Van der Linden PJ, Xiaosu D (2001) Climate Change 2001: The Scientific Basis. Cambridge University Press, Cambridge.

Hughes JB, Daily GC, Ehrlich PR (2002) Conservation of tropical birds in countryside habitats. Ecology Letters 5:121-129

Inbar M, Llerena CA (2000) Erosion processes in high mountain agricultural terraces in Peru. Mountain Research and Development 20:72-79

Kaser G, Osmaston H (2002) Tropical Glaciers. Cambridge University Press, Cambridge

Kintz DB (2003) Land use and land cover change between 1987 and 2001 in the buffer zone of a national park in the tropical Andes. M.A. Thesis, University of Texas at Austin

Kintz DB, Young KR, Crews-Meyer KA (2006) Implications of land use/land cover change in the buffer zone of a national park in the tropical Andes. Environmental Management 38:238-252.

Lambin EF, Geist HJ, Lepers E (2003) Dynamics of land-use and land-cover change in tropical regions. Annual Review of Environment and Resources 28: 205-241

Lambin EF, Turner BL, Geist HJ, Agbola SB, Angelsen A, Bruce JW, Coomes OT, Dirzo R, Fischer G, Folke C, George PS, Homewood K, Imbernon J, Leemans R, Lin X, Moran EF, Mortimore M, Ramakrishnan PS, Richards JF, Skånes H, Steffen W, Stone GD, Svedin U, Veldkamp TA, Vogel C, Xuy J (2001) The causes of land-use and land-cover change: Moving beyond the myths. Global Environmental Change 11:261-269.

Lau GF (2002) Feasting and ancestor veneration at Chinchawas, north highlands of Ancash, Peru. Latin American Antiquity 13:279-304

---------- (2005) Core-periphery relations in the Recuay hinterlands: Economic interaction at Chinchawas, Peru. Antiquity 79: 78-99

Lunt ID, Spooner PG (2005) Using historical ecology to understand patterns of biodiversity in fragmented agricultural landscapes. Journal of Biogeography 32:1859-1873

Mahutga MC (2006) The persistence of structural inequality? A network analysis of international trade, 1965-2000. Social Forces 84: 1863-1889

Mark BG, Seltzer GO (2005) Evaluation of recent glacier recession in the Cordillera Blanca, Peru (AD 1962-1999): Spatial distribution of mass loss and climatic forcing. Quaternary Science Reviews 24: 2265-2280

Maybury-Lewis D (2002) The Politics of Ethnicity: Indigenous Peoples in Latin American States. Harvard University Press and David Rockefeller Center for Latin American Studies, Cambridge

Mayer E (2002) The Articulated Peasant: Household Economies in the Andes. Westview Press, Boulder, Colorado

Mayfield MM, Ackerly D, Daily GC (2006) The diversity and conservation of plant reproductive and dispersal functional traits in human-dominated tropical landscapes. Journal of Ecology 94:522-536

Mayfield MM, Daily GC (2005) Countryside biogeography of neotropical herbaceous and shrubby plants. Ecological Applications 15:423-439

McGarigal K, Marks BJ (1995) FRAGSTATS: A spatial pattern analysis program for quantifying landscape structure. USDA Forest Service. GTR PNW-351

Miller GR, Gill AL (1990) Zooarchaeology at Princay, a Formative period site in highland Ecuador. Journal of Field Archaeology 17:49-68

Myers N, Mittermeier RA, Da Fonseca CG, Gustavo AB, Kent J (2000) Biodiversity hotspots for conservation priorities. Nature 403:853-858

Neel MC, McGarigal K, Cushman SA (2004) Behavior of class-level landscape metrics across gradients of class aggregation and area. Landscape Ecology 19:435-455.

Orlove BS, Chiang JCH, Cane MA (2000) Forecasting Andean rainfall and crop yield from the influence of El Niño on Pleiades visibility. Nature 403:68-71

Perry L, Sandweiss DH, Piperno DR, Rademaker K, Malpass MA, Umire A, de la Vera P (2006) Early maize agriculture and interzonal interaction in southern Peru. Nature 440:76-79

Peterson BJ, McClelland J, Curry R, Holmes RM, Walsh JE, Aagaard K (2006) Trajectory shifts in the Arctic and Subarctic freshwater cycle. Science 313:1061-1066

Postigo J (2006) Change and continuity in human and environment relations in a shepherd community in the high Andes of Peru. M.A. thesis, University of Texas at Austin

Pounds JA, Bustamante MR, Coloma LA, Consuegra JA, Fogden MPL, Foster FN, La Marca E, Masters KL, Merino-Viteri A, Puschendorf R, Ron SR, Sánchez-Azofeifa GA, Still CJ, Young BE (2006) Widespread amphibian extinctions from epidemic disease driven by global warming. Nature 439:161-167

Quincey DJ, Lucas RM, Richardson SD, Glasser NF, Hambrey MJ, Reynolds JM. (2005) Optical remote sensing techniques in high-mountain environments: Application to glacial hazards. Progress in Physical Geography 29:475-505

Radcliffe SA, Laurie N (2006) Culture and development: Taking culture seriously in development for Andean indigenous people. Environment and Planning D: Society and Space 24:231-248

Ramírez E, Francou B, Ribstein P, Descloitres M, Guérin R, Mendoza J, Gallaire R,. Pouyaud B, Jordan E (2001) Small glaciers disappearing in the tropical Andes: A case study in Bolivia: Glaciar Chacaltaya (16°S). Journal of Glaciology 47:187-194

Rhoades GC, Eckert GE, Coleman DC (1998) Effect of pasture trees on soil nitrogen and organic matter: Implications for tropical montane forest restoration. Restoration Ecology 6:262-270

Ricketts TH, Daily GC, Ehrlich PR, Fay JP (2001) Countryside biogeography of moths in a fragmented landscape: Biodiversity in native and agricultural habitats. Conservation Biology 15:378-388

Rosen AM (2006) Civilizing Climate: The Social Impact of Climate Change in the Ancient Near East. AltaMira Press, Lanham, MD

Rudel TK, Richards S (1990) Urbanization, roads, and rural population change in the Ecuadorian Andes. Studies in Comparative International Development 25:73-89

Salazar LF (2006) Emerging and re-emerging potato diseases in the Andes. Potato Research 49:43-47

Salisbury JGT Barnett GA (1999) The world system of international monetary flows: A network analysis. The Information Society 15:31-49

Sandor JA, Eash NS (1995) Ancient agricultural soils in the Andes of southern Peru. Soil Science Society of America Journal 59:170-179

Sarmiento FO (1997) Landscape regeneration by seeds and successional pathways to restore fragile tropandean slopelands. Mountain Research and Development 17:239-252.

Sarmiento FO (2000) Breaking mountain paradigms: Ecological effects on human impacts in man-aged tropandean landscapes. Ambio 29:423-431

Seltzer GO, Hastorf CA (1990) Climatic change and its effect on prehispanic agriculture in the central Peruvian Andes. Journal of Field Archaeology 17:397-414

Silverio W, Jaquet JM (2005) Glacial cover mapping (1987-1996) of the Cordillera Blanca (Peru) using satellite imagery. Remote Sensing of Environment 95:342-350

Stahl PW (1996) Holocene biodiversity: An archaeological perspective from the Americas. Annual Review of Anthropology 25:105-126

Stanish C (1989) Household archeology: Testing models of zonal complementarity in the south central Andes. American Anthropologist N.S. 91: 7-24

Thompson LG, Mosley-Thompson E, Davis ME, Lin PN, Henderson K,. Mashiotta TA (2003) Tropical glacier and ice core evidence of climate change on annual to millennial time scales. Climatic Change 59:137-155

Thompson LG, Mosley-Thompson E, Brecher H, Davis M, León B, Les D, Ping-Nan Lin, Mashiotta T, Mountain K (2006) Abrupt tropical climate change: Past and present. Proceedings of the National Academy of Science 103:10536-10543

Troll C (1968) The cordilleras of the tropical Americas: Aspects of climatic, phytogeographical and agrarian ecology. Colloquium Geographicum 9:15-56

Turner MG (2005) Landscape ecology: What is the state of the science? Annual Review of Ecology and Systematics 36:319-344

Vilimek V, Zapata ML, Klimeš J, Patzelt Z, Santillán N (2005) Influence of glacial retreat on natural hazards of the Palcacocha Lake area, Peru. Landslides 2:107-115

Vuille M, Bradley RS, Keimig F (2000a) Interannual climate variability in the Central Andes and its relation to tropical Pacific and Atlantic forcing. Journal of Geophysical Research 105:12447-12460

Vuille M, Bradley RS, Keimig F (2000b) Climate variability in the Andes of Ecuador and its relation to tropical Pacific and Atlantic sea surface temperature anomalies. Journal of Climate 13:2520-2535

Vuille M, Bradley RS, Werner M, Keimig F (2003) 20th century climate change in the tropical Andes: Observations and model results. Climatic Change 59:75-99

Walker R (2003) Mapping process to pattern in the landscape change of the Amazonian frontier. Annals of the Association of American Geographers 93:376-398

Walsh SJ, Crews-Meyer KA (2002) Linking People, Place, and Policy: A GIScience Approach. Kluwer Academic, Boston

Whitmore TM, Turner BL (2001) Cultivated Landscapes of Middle America on the Eve of Conquest. Oxford University Press, Oxford

Wiegers ES, Hijmans RJ, Hervé D, Fresco LO (1999) Land use intensification and disintensification in the upper Cañete valley, Peru. Human Ecology 27:319-33.

Wulder MA, Dymond CC, White JC, Leckie DG, Carroll AL (2006) Surveying mountain pine beetle damage of forests: A review of remote sensing opportunities. Forest Ecology and Management 221:27-4.

Young KR (1993) National park protection in relation to the ecological zonation of a neighboring human community: An example from northern Peru. Mountain Research and Development 13:267-280

--------- (1997) Wildlife conservation in the cultural landscapes of the central Andes. Landscape and Urban Planning 38:137-147

--------- (1998) Deforestation in landscapes with humid forests in the central Andes: Patterns and processes. In: Zimmerer KS, Young KR (eds) Nature's Geography: New Lessons for Conservation in Developing Countries. University of Wisconsin Press, Madison, pp. 75-99

--------- (2002) Minding the children: Knowledge transfer and the future of sustainable agriculture. Conservation Biology 16:855-856

Young KR, Aspinall R (2006) Kaleidoscoping landscapes, shifting perspectives. The Professional Geographer 58:436-447

Young KR, Ulloa Ulloa C, Luteyn JL, Knapp S (2002) Plant evolution and endemism in Andean South America: An introduction. Botanical Review 68:4-21

Young KR, León B (2007) Tree-line changes along the Andes: Implications of spatial patterns and dynamics. Philosophical Transactions of the Royal Society B: Biological Sciences 362:263-272

Young KR, León B, Jørgensen PM, Ulloa Ulloa C (2007) Tropical and subtropical landscapes of the Andes. In: Veblen TT, Young KR, Orme AR (eds) The Physical Geography of South America. Oxford University Press, Oxford, pp. 200-216.

Young KR, Lipton JK (2006) Adaptive governance and climate change in the tropical highlands of western South America. Climatic Change 78 (1):63-102

Zimmerer KS (1996) Changing Fortunes: Biodiversity and Peasant Livelihood in the Peruvian Andes. University of California Press, Berkeley

---------- (1999) Overlapping patchworks of mountain agriculture in Peru and Bolivia: Toward a regional-global landscape model. Human Ecology 27:135-165

----------(2006) Globalization and New Geographies of Conservation. University of Chicago Press, Chicago

CHAPTER 3 - The Impact of Climate Change on Income Diversification and Food Security in Senegal

Molly E. Brown

Science Systems and Applications, Inc., Biospheric Sciences Branch, Code 614.4, NASA Goddard Space Flight Center, Greenbelt, MD 20771, USA

ABSTRACT

Much uncertainty still exists regarding the impact of global warming and climate change on the semi-arid tropics. This study examines the impact on livelihoods of climate change that has already occurred in the semi-arid region of Senegal, West Africa. Income diversification in response to rainfall reductions over the past 45 years are analyzed using data derived from participatory rural appraisals conducted during the early 1990s. Information on the relative importance of income sources from 23 rural communities is presented for the past, present and projected changes in the future. Multiple correspondence analysis is used to provide a quantitative evaluation of how different levels of rainfall deficits affect diversification of income away from agriculture, and the resulting ability of communities to access food throughout the year. The research shows that in regions that have experienced large reductions in rainfall since 1951, income diversification has been significant, suggesting possible future policy pathways for the future.

INTRODUCTION

Climate change models disagree as to the impact of increasing greenhouse gases and climate change on the arid and semi-arid zones of West Africa (van den Born et al. 2004; Stainforth et al. 2005). Some models predict significant temperature increases and simultaneous declines in rainfall, others show either an ambiguous result or an improvement in precipitation (Xue and Shukla 1993; Janicot 1994; Clark et al. 2001; Maynard et al. 2002). Even a small reduction in rainfall will cause a change in the start and end of the growing season, resulting in possibly large disruptions in the existing agricultural systems (Verhagen et al. 2004). During the past 30 years, many areas of West Africa have experienced declines in rainfall that approximate these predicted changes (Dai et al. 2004). Documented reductions in forest species ranges and changes in cropping regimes have been documented as a response to these reductions in rainfall (Haggblade et al. 1989; Tucker et al. 1991; Gonzalez 2001; Nicholson 2001; Taylor et al. 2002). In this study, we examine the socio-economic adjustments to climate change that has already occurred in the semi-arid zone of Senegal as a model for possible responses to climate change that may occur elsewhere in the future.

In this case study, variable rainfall trends over 45 years are compared with stakeholder strategies in a homogeneous rural livelihood and agro-ecological zone. Using socio-economic data gathered from participatory rural appraisals (PRAs), income diversification, food security, and trends in rainfall are examined. Previous work on climate variability and income diversification in Burkina Faso by Reardon et al. (1992) showed that reductions in agricultural production are a factor in household income diversification, and that non-farm income is used to compensate for these shortfalls.

The data used in this chapter is derived from reports from participatory rural appraisals conducted by a variety of non-governmental organizations during the 1990s provides both a long-term perspective and sufficiently regional coverage to determine how strategies change with changing rainfall trends (Figure 3.1). Although the socio-economic data that are analyzed here are derived from a qualitative source whose relationship to beliefs and strategies in the region as a whole cannot be determined, they do provide the opportunity to conduct analysis to enlighten the interaction between climate, development and diversification of income sources.

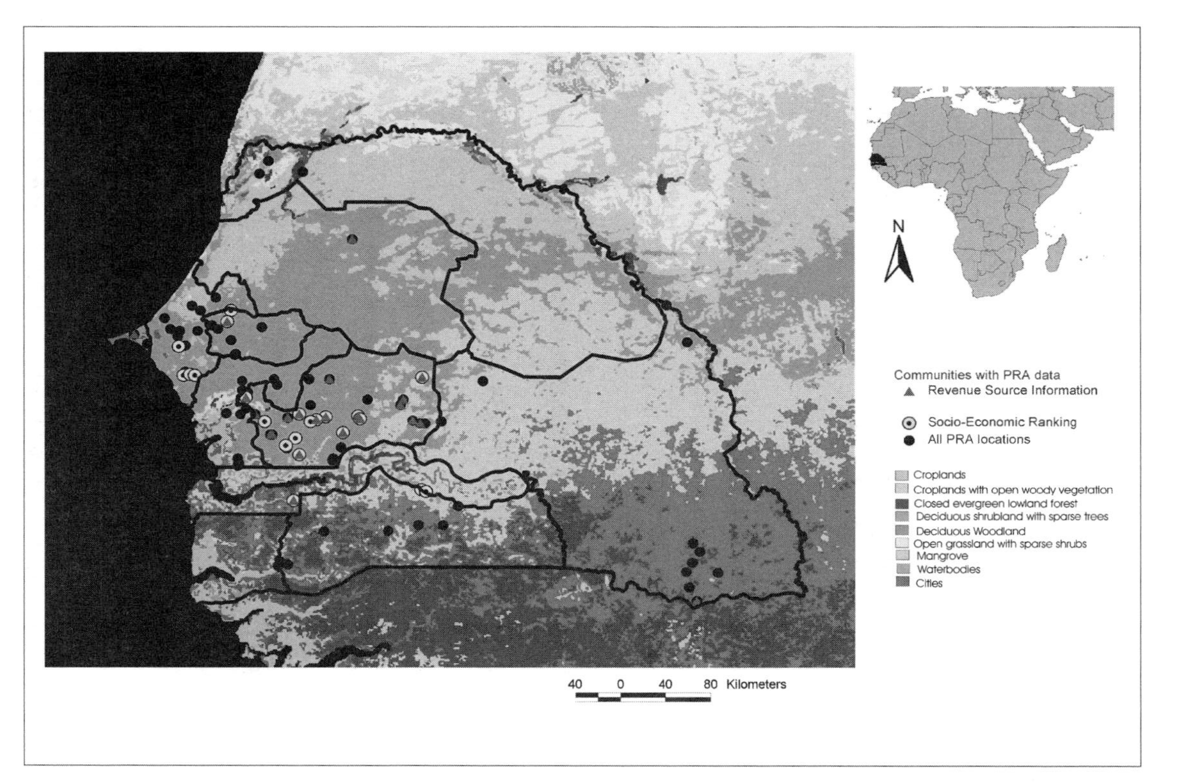

Fig.3.1. Map of communities where PRA were conducted, overlaid on land-use map. Communities with information on revenue sources and poverty levels are identified. Base map shows the land use/land cover information from regional Africa dataset Global Land Cover 2000 (Mayaux et al. 2004)

The data used in this research would not support generalizations of the sort that Ellis (1998) mentions, but does provide the opportunity to examine how rainfall, a very important production determinant in West Africa, co-varies with livelihood strategy. Such an analysis will provide insight into how the region may respond to future changes in climate, resulting in systematic reductions in rainfall.

The paper's objectives are to show how long-term reductions in rainfall in specific localities of Senegal have affected income diversification away from rain-fed agriculture, how these activities are associated with natural resource management problems, and how income diversification affects food security for the most vulnerable. Senegal has already experienced a significant change in their climate during the past 45 years; therefore, changes in livelihood strategies should guide policies and programs that aim at improving adaptation to climate change by rural agricultural societies in regions that may experience significant reductions in rainfall during the next century.

DATA

Participatory Rural Appraisal Documents as a Data Source

Heuristics for conducting PRAs generally follow the approach documented by McCracken et al. (1988): "A semi-structured activity carried out in the field, by a multi-disciplinary team and designed to acquire quickly information on, and new hypotheses about, rural life." A review of methods of poverty mapping by Davis (2003) compares participatory methods with more formal surveys and census approaches in effort to estimate poverty and income levels in lesser-developed countries. Although participatory approaches are of necessity subjective and difficult to quantify, they have an advantage over census income variables that tend to be biased in regions where a high share of income is informal or non-cash in nature (Davis 2003). Indicators of wealth and income are self-identified by the people themselves and are thus less prone to bias (Boudreau 1998). Despite the fact that the data are uncertain and do not represent a statistically valid sample of the diversity of economic livelihoods in Senegal, they are still interesting for examining the potential responses of the communities in the future if rainfall were to continue to decline in the manner in which it has during the past 45 years.

The locations of all communities in Senegal where documents from PRAs have been collected are shown in Figure 3.1, along with the subset communities reporting information on revenue sources and socio-economic ranking. The community appraisals were conducted during the 1990s by ten non-governmental organizations and collected by the author in 1999 during a visit to Senegal. Table 3.1 shows the sponsoring organizations that conducted or commissioned the subset of reports used in this study.

Table 3.1. Sources of Participatory Rural Appraisal data used in this study

Total PRAs	Revenue subset	Sponsoring organization
26	12	Africare – Senegal
21	1	USAID Natural Resources Project (PGCRN)
20	0	Projet D'Appui/USU – ACDI – USAID projects
8	4	Development Assistance for NGOs (CONGAD)
6	1	Land Tenure Center, Univ of Wisconsin
6	1	Rodale Institute, Senegal
5	0	World Vision International, Senegal
4	2	International Institute for Environment and Development (IIED)
2	1	Village reforestation in the north-west groundnut basin project (PREVINOBA)
2	1	Action Aide The Gambia

In Senegal, the US Agency for International Development-funded programs have centralized the training of PRA practitioners. There are 23 documents that have information on past, present and future income sources from documents written and supplied by seven non-governmental and governmental organizations, 22 documents have information about the socio-economic classes of the communities, and 11 documents have both revenue and socio-economic information. This has resulted in a similar methodology and practice of PRA across regions and sponsoring organizations, despite different objectives for the PRAs (USAID 1997; Africare 1998; Brown 2006)

Data on Revenue Sources

The revenue data presented in the PRA reports were collected using semi-structured interviews and reported in 23 PRA reports. Revenue data were collected using informal interviews with a diverse set of community members. Groups of women and men – young and old – were interviewed and asked about their sources of income. Each group of people interviewed were asked to report income sources in the past, current sources of income and to project changes in the future. In many cases, the entire dataset generated by the interviews was reported in the PRA document, as well as summarized in the form of a table or pie chart in the text. The tool was used to summarize the income structure of the community, which required a consensus to be reached by all those interviewed. The data were grouped into seven income categories in order to compare directly the revenue sources from different communities. Data from men and women were combined to give an overall picture of the revenue sources for the community (Table 3.2).

The categories in Table 3.2 were chosen to focus on the division between activities that would be directly affected by variations in rainfall (agriculture, livestock, and to a smaller extent, gardening) and those that are less affected (commerce, services, gathering, and remittances), but still vulnerable to a critical reduction in economic activity that may occur during drought years. These are also referred to as the farm and non-farm sectors in the literature (Ellis 1998; Bryceson 2002). Because the data are qualitative and the types of activities that produce income were self-specified by the participants themselves instead of set by the researchers, a further specification of these categories is not possible.

Data on Problems Reported by Communities

To compare the problems between diverse communities, reports, and assessment objectives, a binary variable was applied (Gonzalez 2001). If a PRA report mentioned a problem in a community, it was given a one; otherwise it was given a zero. By using a yes/no presence test, all problems mentioned in the PRA report can be recorded, creating a semi-quantitative variable from the information presented for each community. Eight management problems are included here (Table 3.3) and describe socio-economic and natural resource management (NRM) issues (Brown 2006). These problems were selected as those most directly related to agricultural and income diversity issues, and are used to elaborate on the livelihood strategies of communities diversifying their income. If the problem was mentioned in the text of the document as an issue in the community, the problem was given a one, if it was not mentioned, then it was given a zero.

Table 3.2. Descriptions of seven income source categories created to summarize income from 23 communities in Senegal

Category	Description	Past	Present	Future
Agriculture	Rainfed agricultural cash and subsistence crops, including millet, peanuts, fonio, maize, manioc, cow pea, and bessy	51%	39%	38%
Gardening	Fruit and vegetables grown in protected areas, often requiring irrigation, including tomatoes, onions watermelons, okra, traditional Senegalese vegetables of diakhatou, gadianga and nadie, and cultivated fruit crops such as mango.	13%	12%	8%
Livestock	Purchasing, raising and feeding of domestic animals for sale, and for dairy and other products to be sold on the market.	12%	15%	20%
Commerce	Selling of products in markets, including the creating and marketing food products such as donuts, selling of cloth, batteries, tea, coffee, and sugar in the community and in markets, and renewable animal products such as milk and yogurt. This category also included the purchase and resale of goods across borders.	10%	18%	20%
Services	Masonry, carpentry, hair braiding, hair cutting, hand decoration using henna, traditional crafts and artisinal products such those purchased by tourists, ferrying goods using animal drawn carts, engine repair, car/bus driving, and income generated by working in the national transport system.	8%	8%	9%
Gathering	Wild foods and timber products gathered freely from open -access land, including income from the collection and marketing of salt, condiments made from wild tree products, firewood, charcoal, building materials such as thatch for roofing and poles, fence materials, and fishing.	6%	5%	3%
Remittances	Income from family members living in nearby villages, towns, cities and abroad	0%	3%	2%

Table 3.3. Description of eight natural resource management problems extracted from the text of the PRA documents

Problem	Description
Soil	Soil fertility of agricultural fields reduced significantly than was remembered in the recent past. Erosion problems from water and wind, including gullying.
Migration	Migration of community members mentioned in report.
Deforestation	Deforestation is mentioned as a significant problem. Investments in reforestation activities such as planting trees, protecting natural regeneration are mentioned. Technical assistance is often requested by the village.
Labor	Reduced manpower for agriculture and other activities is a significant issue in managing natural resources.
Revenue	Inadequate revenue is generated or food grown throughout the year, causing villagers to spend significant amounts of time and effort to find additional sources of food or revenue during bad years or annually during the summer months (*soudure*).
Yields	Reduced yields due to reduced agricultural inputs and reduced rainfall are mentioned as an important contribution to reduced income for entire community.
Land	Increased population and reduced yields have increased land requirements above that which are available to the village. This has led to land scarcity which significantly reduces the ability of the population to meet its needs by farming.
Credit	A lack of credit to purchase farm implements, seeds and fertilizer is mentioned as a significant source of reduced revenue generating activity.

Precipitation Data

The long-term slope trend in rainfall from 1951 to 1995 was obtained from the Climatic Research Unit[1] , and is a 3-minute resolution gridded rainfall dataset for Africa created from 2307 precipitation stations throughout Africa (New and Hulme 1997, MARA/ARMA 2004). The data is gridded using thin-plate splines to interpolate the mean climate surfaces as a function of latitude, longitude and elevation (Wahba 1979; Hutchinson 1995). Beginning in 1951, the data provides a historical perspective of how rainfall has changed through time (Figure 3.2).

[1] Located at the University of East Anglia, http://www.cru.uea.ac.uk/

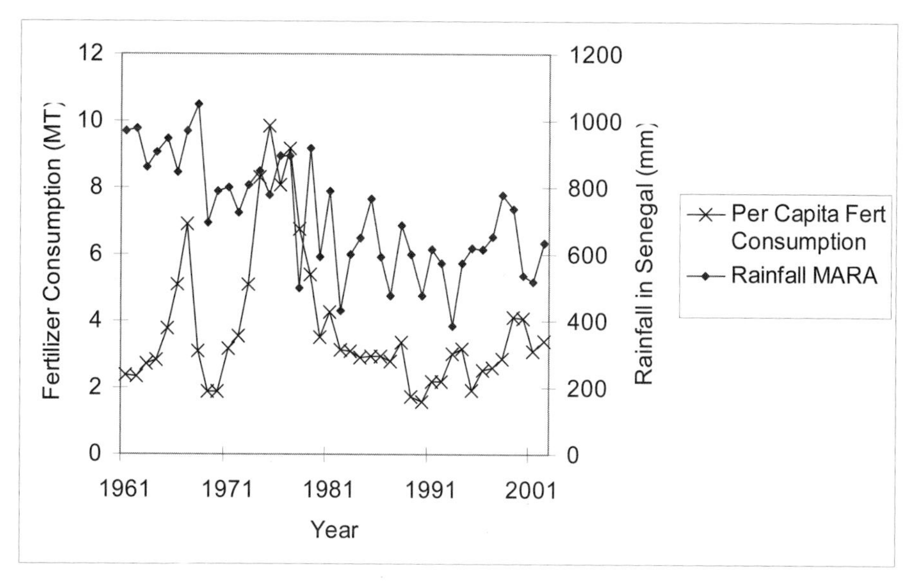

Fig. 3.2. Per capita fertilizer production and average rainfall over Senegal from the Mapping Malaria Risk in Africa (MARA) project

This length of record is coincident with many community members' perceptions of the change in the climate during their lifetimes, and enables the grouping of communities with similar reductions in precipitation together to show how these changes have affected livelihood strategies. The inter-annual variability of rainfall from 1951 to 1995 was measured using the standard deviation of the de-trended annual rainfall totals. This measure of variability provides information about how reliable agricultural production is while normalizing for the overall precipitation level.

METHODOLOGY

Beyond presenting the information collected from the PRA, this paper seeks to relate these data to changes in the environmental situation of the communities. This is done using multiple correspondence analysis (MCA), a methodology well suited to the quasi-quantitative variables that have been constructed from the PRA reports (Greenacre 1984). MCA analysis is conducted on categorical matrices that are generated by counting the frequencies of the occurrences of the different variables in relationship to each other. First the technique will be described and then details on how the categorical matrices were constructed will be presented.

Correspondence Analysis

MCA is a technique for displaying associations among a set of categorical variables in a scatterplot or map, that allows a visual display of the patterns within the data (Everitt and Dunn 2001). The analysis indicates whether certain levels of one trait are associated with some levels of another. Conducting a geometric analysis on the resulting two-way contingency table summarizes the observed association of the two traits. Our implementation of correspondence analysis uses the singular value decomposition of the matrix whose elements are based on the chi-squared statistic. Details of this implementation can be found in Everitt and Dunn (2001). The MCA map can be interpreted by comparing the distances between the row points and the distances between the column points in order to show association between the variables. When two variables are close together on the MCA map, they are more closely associated with one another than two variables that are farrther away (Everitt and Dunn 2001; Greenacre 1984).

Categorical Matrices

The categorical matrices used here are two-way tables that present the data by counting the number of observations that fall into each group for two variables, one divided into rows and the other divided into columns. The first categorical matrix decomposed was based on the natural resource management problems reported by the communities and the sources of income. First, the communities experiencing changes in the different income sources were found. Then, in this subset of communities, the number that reported the presence of a problem are counted and this frequency reported in the matrix. This creates a matrix of seven rows (income sources) by eight columns (the presence of NRM problems). A similar matrix was created between the eight NRM problems and the communities divided into four categories of food insecurity, as reported by the PRA socio-economic ranking exercise.

The second set of correspondence plots show the relationships between rainfall change from 1951 to 1995 in millimeters of rainfall per year, the inter-annual rainfall variability and income diversity change through time. Diversity is measured by the change from the past to the present of the number of categories a community derives its income from. This metric varies from -1 (a reduction of income sources by one category) to +3 (an increase of three income categories). By finding the communities with a change in categories (-1 to +3), and then counting the number of communities within that subset that had a small (-4 to -7 mm/year), medium (-7 to -

9 mm/year) or large (-9 to -12 mm/year) reduction in rainfall, a matrix of three rows by five columns was created. A similar matrix using income diversity and rainfall variability was created by finding the communities with five levels of change in income (-1 to +3) and then counting the number of communities with different levels of rainfall variability in the subset of communities. The resulting four rows by five columns matrix was created and decomposed using MCA.

RESULTS

As the principal source of income in rural communities, agriculture has declined in its importance in relation to other sources of income. This finding was similar to that of Reardon et al. (1992) and more recently Bryceson (2002), in which declining world commodity prices and reduced subsidies for agricultural inputs, particularly fertilizer, have put pressure on agricultural revenues for subsistence and cash cropping farmers throughout Africa. Figure 3.3 shows a histogram of the revenue streams from Senegal as reported in the PRA documents. Increasing reliance on commerce has offset lost agricultural and livestock revenue sources, with stable contributions by wild food gathering, irrigated gardening, service provision and remittances from migrants.

Information on revenue sources also revealed a slight decline in the importance of gathering and marketing wild products by communities. Trends in wild product marketing reflects the declining availability of wild resources that can be accessed by anyone in the community, due to both land-use change and the increased harvesting and transformation of these products into marketable products. Community members describe how disappearing wild fruit tree species and local vegetative building materials that used to be in abundance have forced greater reliance on the market for these products. Expansion of agriculture into previously uncultivated areas is cited as one reason for these changes, although the significant reduction in rainfall over the past 45 years has reduced the viability of many of the tree species in this semi-arid region (Gonzalez 2001). The PRAs also document a drop in the availability of natural materials needed to build houses, fences, and to produce wild-product based foods. These products must now be purchased in the marketplace, increasing the demand for cash resources, which in turn forces more activities that generate cash income, further increasing household exposure to the marketplace.

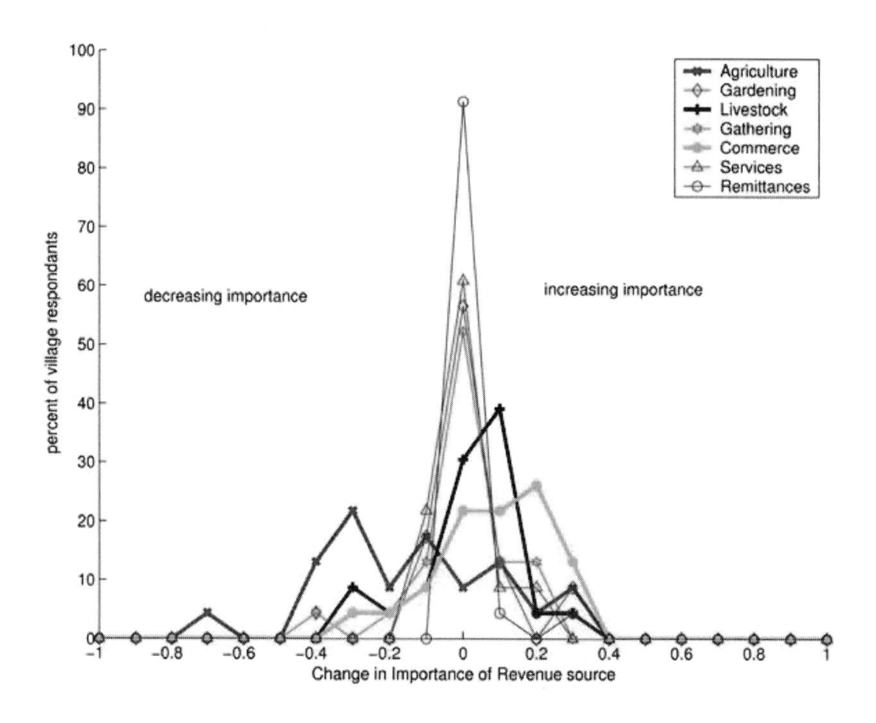

Fig. 3.3. Histogram of percent change of revenue sources from the past to the present by category. Units of the x axis are in change of income as a percent of the total.

Natural Resource Problems Associated with Revenue Generation

To evaluate the socio-economic and environmental problems that rural Senegalese communities have experienced, a correspondence analysis was conducted to show the relationships between natural resource management problems and revenue trends. The problems most closely associated with declining agriculture and increasing livestock income are insufficient revenue, declining yields, and deforestation (Figure 3.4). Reductions in agricultural productivity are closely tied to the shift in emphasis from rainfed agriculture consisting of millet, peanut and legume cropping to investment in livestock as both a source of manure as a fertilizer and as a source of income. Remittances, migration and declining agricultural income are closely associated with one another.

Migration is a common response to changes in the productivity of agriculture, hence the close association of remittances with agriculture and its associated problems (Glantz 1987; Golan 1994). In Figure 3.4 migration refers to the presence/absence test for migration in the community, and was collected using the text of the PRA reports instead of the income matrix tool. The closer the problems and revenue sources are, the more closely the correspondence between the variables is. The variance explained by this correspondence map is 40.7 percent for the first dimension and 65.2 percent in the second dimension.

In some communities, migration has become routine. For example, in Nguick Fall, Arrondissement Pambal in the Thiés Region, 9.4 percent of the population (73 people) went on seasonal migration in 1996, the year the PRA was conducted. They went to Dakar, other cities in the regional, and local towns to work as chauffeurs, apprentice chauffeurs, domestic labor, students and traders. Some return every year for the growing season, some only return for holidays and family visits. Despite this, 92.5 percent of the population report being primarily agriculturalists, although half of this activity now involves irrigated gardening instead of rainfed agriculture. This increase in irrigation has become possible due to significant investments in water extraction machinery, the communities' location over a healthy aquifer, and access to a road leading to a large urban market. Declining yields due to inadequate soil augmentation means that in an average year, residents report that their peanut fields produce an average 250-350 kg ha^{-1}, down from >700 kg ha^{-1} several decades before (Figure 3.2). Organic fertilizer and fallow are used to improve the soil wherever possible, thereby increasing the value of livestock in the community, but these inputs have not offset the reduction in inorganic fertilizer use during the past five decades.

Other results from Figure 3.4 show that gardening and collection of wild products, both labor-intensive activities, are associated with the problems of insufficient labor and scarcity of land. Migration and the provision of services as income sources are associated with each other, as often migrants learn and use skills that can then be used to generate income locally. Commerce (working in the marketplace) is on the rise nearly everywhere, which may be due to reductions in soil fertility (present in 87 percent of communities with PRA reports), as the two are associated with one another (also see Figure 3.2).

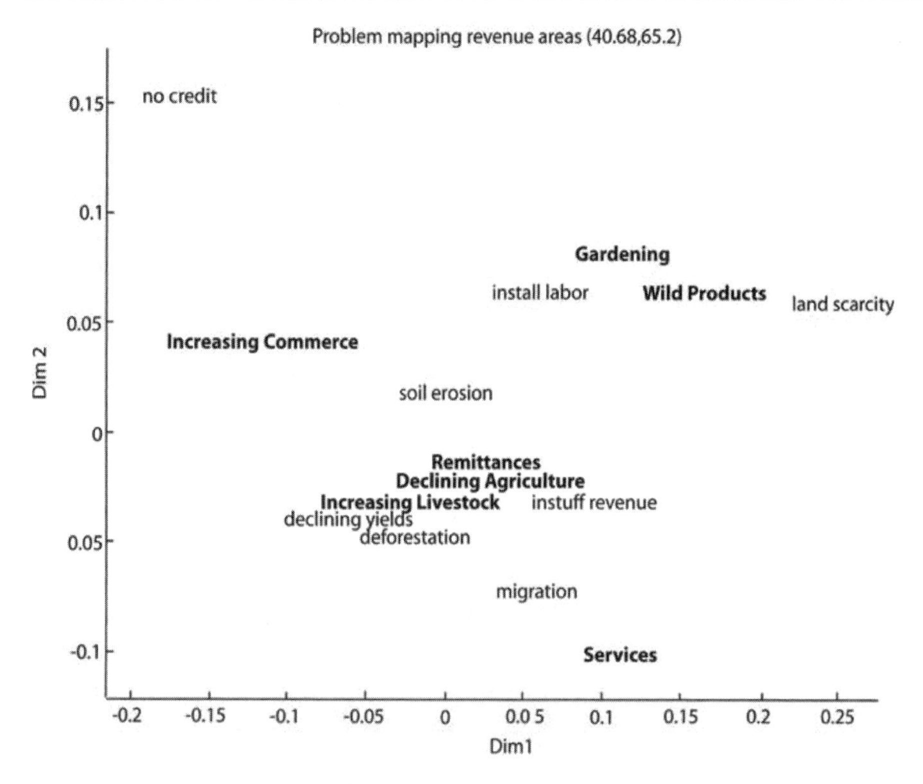

Fig. 3.4. Correspondence of changing revenue sources and natural resource problems presented in Table 3.2 found in the 23 communities

DIVERSITY OF REVENUE AND LONG TERM CHANGES IN RAINFALL

The question of how income diversity changes in relationship to long-term trends in rainfall is explored in Figures 3.5 and 3.6. The changes in diversity of income sources from the past to the present were related to reductions in rainfall during the period of record. Large reductions in rainfall are closely associated with increases in income sources, most frequently the addition of commerce and livestock to supplement decreasing income from agriculture. Moderate changes of rainfall are associated with negligible change in income diversity. Finally, regions with only small reductions in rainfall during the past four decades have experienced no overall change in income sources according to this data. It is expected that smaller overall rainfall totals will result in lower agricultural income, larger income diversification should be associated with larger changes in rainfall, which is seen in Figure 3.5.

Rainfall variability was also associated with increases in diversity. Communities experiencing very high levels of variability experienced larger rates of income diversification, compared to communities with lower rainfall variability. Moderate rainfall variability was associated with communities with increases of three categories of income from which they derived their income.These results support the findings of Golan (1994), who showed that the level of income diversification depended on the economic and political ties of its community members, as well as their increasing needs for new sources of income. In regions farther north that have fewer agricultural options due to much less rainfall overall, diversity has increased more consistently than in regions that have experienced less severe declines in rainfall.

In Figure 3.6 income diversity is a measure of the number of sources of income created by subtracting the number of income categories that a community reported now from the past. The diversity is described both in words and in numbers: -1 for a decline in one category, 0 for no change, +1 for an increase of one category.

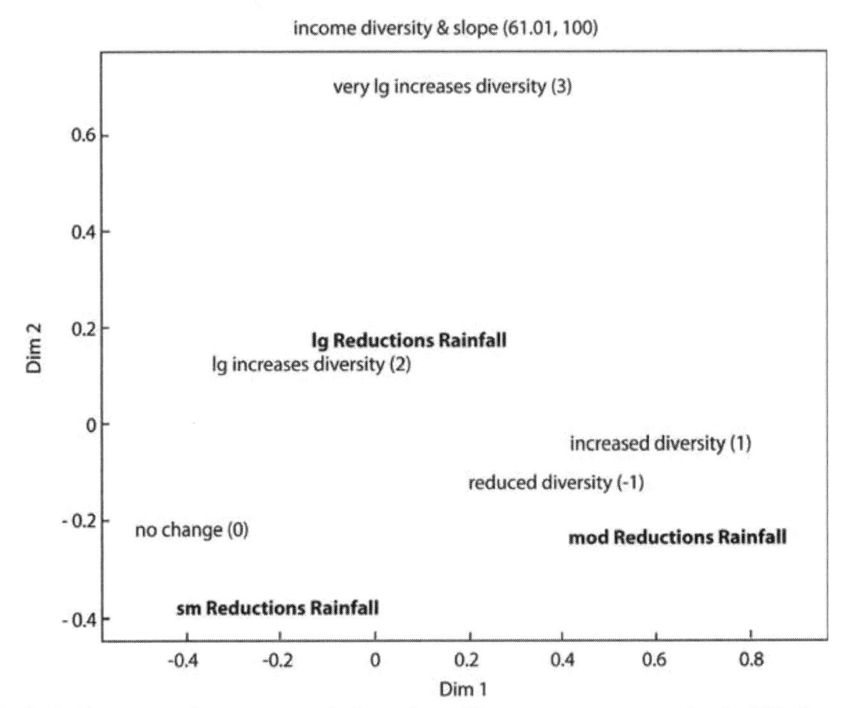

Fig.3.5. Correspondence map of diversity of income sources and rainfall changes through time

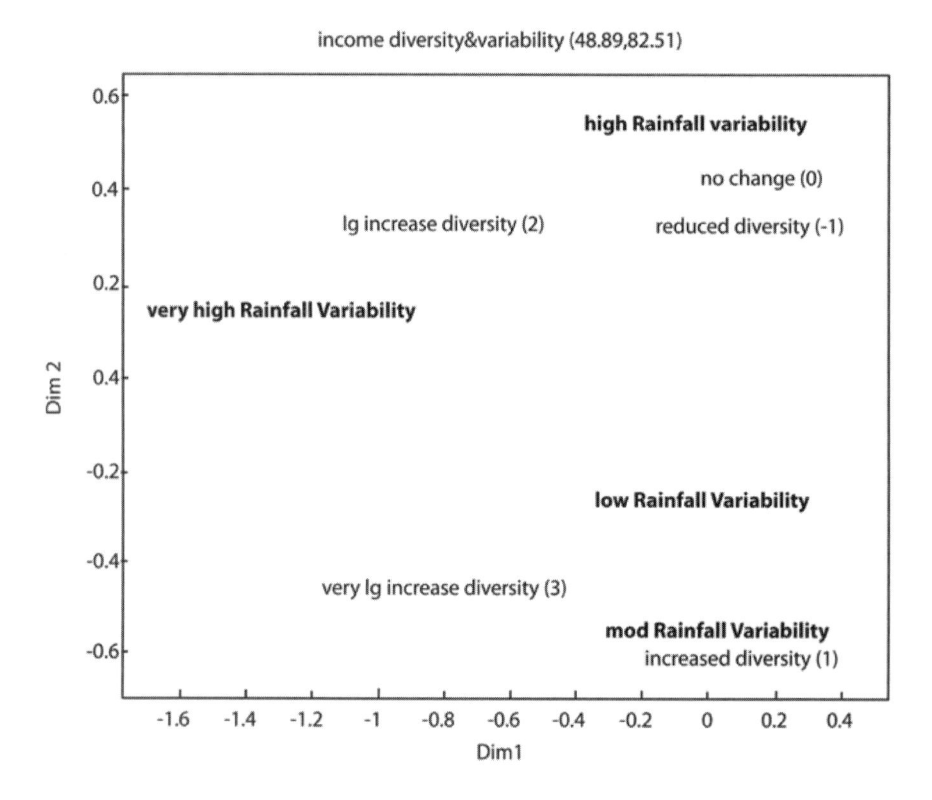

Fig. 3.6. Correspondence map of diversity of income sources and rainfall variability. The first dimension explains 55 percent of the variance and with the first dimension and the second dimension explains 100 percent of the variance

DISCUSSION AND CONCLUSION

Senegal's experience in adapting to changing rainfall levels over the past half century has many policy and programmatic implications for organizations wishing to prepare the semi-arid agricultural regions for possible climate change in the coming century.

To maintain their livelihoods in the face of declining rainfall, most communities have diversified their income sources away from agriculture into many other activities. Scarcity of rainfall has impacted not only agriculture, but also the availability of natural resources for the most vulnerable groups. The poor, who are heavily dependent on open access natural resources, are the most severely affected by deteriorating environmental conditions and factors limiting resource access (Gregory et al. 2005). While climate change is not the only threat to natural resources, climate-induced changes to resource flows will affect the viability of livelihoods unless measures are taken to protect and diversify them through programs and policies that focus on economic development outside of agriculture.

Senegalese agriculturalists see falling income and reduced yields as a trend that will continue into the future. Because agricultural incomes may not improve in the next decade, agriculturalists will encourage their children to move into commerce, service jobs, and out-migrate to cities. Many communities want to diversify their traditional cultivation system based on millet and peanuts with irrigated gardening and raising livestock. Fresh vegetables and dairy products are marketable in the growing towns and cities in a way that traditional grains are not. These activities will need investments in securing water sources and machinery to harvest water, in agricultural equipment, in seeds, and in the skills and knowledge required to be successful in these enterprises. These barriers to diversifying income and agricultural activities are difficult to overcome and are often the target of development projects.

The analysis presented in this paper has shown that diversification of livelihoods has been most successful in regions with the largest changes in rainfall, but that these regions also experience relatively large amounts of food insecurity. Those with the largest changes in income sources have experienced only relatively modest amounts of rainfall decline, suggesting that income diversification requires sufficient capital that is often derived from agriculture during years with above-average rainfall. When rainfall is so low that it falls below the minimum to support peanuts or millet then diversification of income sources becomes more difficult. Similar conclusions were drawn by Reardon et al. (1992).

Policies and programs should focus on improving access to new sources of income, such as irrigated market gardening of products in demand in the growing urban areas, such as fresh fruit and vegetables, cultivated wild food products and culturally-important ceremonial goods. These products will be marketable both in urban areas and overseas to the growing African diaspora. With increasing globalization, a focus on markets and products is key to any communities' success. Successful gardening requires access to water and to the skills needed to maintain machinery for extracting and distributing it. Education to effectively manage tgardens and the revenue from this type of enterprise are urgently needed in order to support a growing population. Development of other sources of income can also be supported through education and outreach programs focused on providing education and training for the many young people looking for new ways of supporting their families in the countryside. Effective policies that encourage income diversification and reduce pressures on land may also contribute to the continued viability of agriculture in the region. Without sufficient rainfall, agriculture and even gardening income levels will continue to drop.

Analysis of revenue and resource data found in PRA reports supplemented with rainfall data support the conclusion that diversification from a primarily agricultural income source occurs in response to external forces such as climate change. This diversification is manifested by an increase in livestock and commercial income activities as per capita agricultural income declines. Reliance on agricultural income is further reduced by a growing trend of migration out of areas experiencing food insecurity. Application of multiple correspondence analysis to data extracted from participatory rural appraisals is effective at displaying the association between rainfall deficits, income diversification and food security. Lessons that can be recommended from livelihood strategy change in response to the observed climate change include policies and programs that support and encourage adaptation to new economic opportunities, especially those that support the population in the growing urban areas in Senegal.

ACKNOWLEDGEMENTS

The author would like to thank the University of Maryland Committee for the African Diaspora for a grant that permitted the collection of the documents used in this paper. She would also like to thank Pierre Defourny, Devin Paden, David Rain and Dan Slayback for comments on earlier versions of this paper.

REFERENCES

Africare (1998) Agricultural Enterprise and Health Development Program - Fiscal Years 1999-2001, Africare, Washington DC

Boudreau TE (1998) The Food Economy Approach: A Framework for Understanding Rural Livelihoods. Report No. RRN Network Paper 26, Relief and Rehabilitation Network/Overseas Development Institute, London

Brown ME (2006) Assessing Natural Resource Management Challenges in Senegal Using Data from Participatory Rural Appraisals and Remote Sensing. World Development 34:751-767

Bryceson DF (2002) The Scramble in Africa: Reorienting Rural Livelihoods. World Development 30:725-739

Clark DB, Xue Y, Harding RJ, Valdes PJ (2001) Modeling the impact of land surface degradation on the climate of Tropical North Africa. Journal of Climate 14:1809-1822

Dai A, Lamb PL, Trenberth KE, Hulme D, Jones PD, Xie P (2004) Comment: The Recent Sahel Drought is Real. International Journal of Climatology 24:1323-1331

Davis B (2003) Choosing a Method for Poverty Mapping, Food and Agriculture Organization of the United Nations, Rome

Ellis F (1988) Peasant Economics: Farm Households and Agrarian Development, Cambridge University Press, Cambridge

Ellis F (1998) Household Strategies and Rural Livelihood Diversification. The Journal of Development Studies 35:1-38

Everitt BS, Dunn G (2001) Applied Multivariate Data Analysis, 2nd edn. Oxford University Press, New York

Glantz M (1987) Drought and Economic Development in sub-Saharan Africa. In: Glantz M (ed) Drought and Hunger in Africa: Denying Famine a Future. Cambridge University Press, Cambridge

Golan EH (1994) Sustainability and Migration Experiments from the Senegalese Peanut Basin. Annals of Regional Science 28:91-106

Gonzalez P (2001) Desertification and a Shift of Forest Species in the West African Sahel. Climate Research 17:217-228

Greenacre MJ (1984) Theory and Applications of Correspondence Analysis. Academic Press, London.

Gregory PJ, Ingram JS, Brklacich M (2005) Climate Change and Food Security. Philosophical Transactions of the Royal Society B 360:2139-2148

Haggblade S, Hazell P, Brown J (1989) Farm-Nonfarm Linkages in Rural Sub-Saharan Africa. World Development 17:1173-1201

Hutchinson MF (1995) Interpolating Mean Rainfall using Thin Plate Smoothing Splines. International Journal of Geographic Information Systems 9:385-403

Janicot S (1994) The Recent West African Rainfall Variability through Empirical and Modelling investigations. In: Desbois M, Desalmand F (eds) Global Precipitations and Climate Change, Vol. I, Series 26, NATO ASI, Berlin, pp 135-149

MARA/ARMA (2004) Monthly Rainfall and Temperature Surfaces for Africa 1951 - 1995 (CD-ROM). Climatic Research Unit

Mayaux P, Bartholome E, Fritz S, Belward AS (2004) A New Land Cover Map of Africa for the Year 2000. Journal of Biogeography 31:861-877

Maynard K, Royer JF, Chauvin F (2002) Impact of Greenhouse Warming on the West African Summer monsoon. Climate Dynamics 19:499-514

McCracken J, Pretty J and Conway G (1988) An Introduction to Rapid Rural Appraisal for Agricultural Development. IIED, London.

New M, Hulme D (1997) Monthly Rainfall and Temperature Surfaces for Africa 1951 - 1995 (CD-ROM), University of East Anglia, Norwich, UK

Nicholson SE (2001) Climatic and environmental change in Africa during the last two centuries. Climate Research 17:123-144

Reardon T, Delgado C, Matlon P (1992) Determinants and Effects of Income Diversification Amongst Farm Households in Burkina Faso. The Journal of Development Studies 28:264-296
Stainforth DA, Aina T, Christensen C, Collins M, Faull N, Frame DJ, Kettlevborough JA, Knight S, Martin A, Murphy JM, Piani C, Sexton D, Smith LA, Spicer RA, Thorpe AJ, Allen MR (2005) Uncertainty in Predictions of the Climate Response to Rising Levels of Greenhouse Gases. Nature 433:403-406
Taylor JE, Lambin EF, Stephenne N, Harding RJ, Essery RLH (2002) The influence of Land Use Change on Climate in the Sahel. Journal of Climate 15:3615-3629
Tucker CJ, Dregne HE, Newcomb WW (1991) Expansion and Contraction of the Sahara Desert from 1980 to 1990. Science 253:299-301
USAID (1997) USAID Technical Evaluation of Africare's Kaolack Agricultural Enterprise Development Program, USAID-Senegal, Dakar, Senegal
van den Born GJ, Leemans R, Schaeffer M (2004) Climate change scenarios for drylands West Africa, 1990-2050. In: Dietz AJ, Ruben R, Verhagen A (eds) The Impact of Climate Change on Drylands: With a Focus on West Africa, Vol 39. Kluwer Academic Publishers, Dordrecht, The Netherlands, pp 43-47
Verhagen J, Put M, Zaal F, van Keulen H (2004) Climate change and Drought Risks for Agriculture. In: Dietz AJ, Ruben R, Verhagen A (eds) The Impact of Climate Change on Drylands: With a Focus on West Africa. Kluwer Academic Publishers, Dordrecht, The Netherlands, pp 49-59
Wahba G (1979) How to smooth curves and surfaces with splines and cross-validation 24th Conference on the Design of Experiments. U.S. Army Research Office, pp 167-192
Xue Y, Shukla J (1993) The Influence of Land Surface Properties on Sahel Climate. Part 1: Desertification. Journal of Climate 6:2232-2245

CHAPTER 4 - Land-Use Changes and Agricultural Growth in India and Pakistan, 1901-2004

Takashi Kurosaki

Institute of Economic Research, Hitotsubashi University, 2-1 Naka, Kunitachi, Tokyo 186-8603 Japan

ABSTRACT

As a case study of globally important agricultural landscapes in the tropics, this chapter investigates changes in land use in the Indo-Gangetic Plains. Although these regions experienced substantial changes in land use and a rapid agricultural growth in the twentieth century, the absolute income level of farmers still remained at the level of low-income countries and the number of the absolute poor was the largest in the world at the end of the twentieth century. With this background, this chapter describes the land-use changes in India and Pakistan, associates the changes with long-term agricultural performance, and shows the importance of crop shifts in enhancing aggregate land productivity, which is a source of growth unnoticed in the existing literature. The use of unusually long-term data that correspond to the current borders of India and Pakistan for the period 1901-2004 also distinguishes this study from the existing ones. The growth records of agricultural production and shifts in crop mix indices show that changes in aggregate land productivity were associated structurally with inter-crop and inter-district reallocations of land use. These changes reflected comparative advantage and contributed to the improvement of aggregate productivity. The crop concentration indices were at their highest levels in the early 2000s both in India and in Pakistan, showing the effects of agricultural liberalization policies and farmers' response to these policies.

INTRODUCTION

To halve the proportion of people whose income is less than one dollar a day and to halve the proportion of people who suffer from hunger between 1990 and 2015 are the first two targets of the Millennium Development Goals. Whether these targets will be achieved critically depends on the performance of the South Asian region where the number of the absolute poor is the largest in the world. According to World Bank (2001), the number of people living on less than one dollar a day in 1998 was 522 million in South Asia, out of the global total of 1.199 billion. At the same time, the two largest countries in the region, India and Pakistan, experienced a rapid agricultural production growth in the twentieth century. How was the growth achieved? Why was the growth not sufficient to substantially reduce the number of the poor? These are questions that motivated the research that is reported in this chapter to investigate changes in land use in the Indo-Gangetic Plains during the last century.

This chapter describes the land-use changes in India and Pakistan, associates the changes with long-term agricultural performance, and shows the importance of crop shifts in enhancing aggregate land productivity, which is a source of growth unnoticed in the existing literature.[1] The use of unusually long-term data that correspond to the current borders of India and Pakistan for the period 1901-2004[2] also distinguishes this study from the existing ones on long-term agricultural development in the Indo-Gangetic Plains.[3] This chapter is distinguished from other chapters in this book by its focus on analyzing statistical records using a new decomposition method proposed by Kurosaki (2003).

The chapter is organized as follows. The next section defines the spatial coverage of the analyses and describes long-term changes in land utiliza-

---

[1] Historical records show that agricultural productivity has increased thanks to the introduction of modern technologies, the commercialization of agriculture, capital deepening, and factor shifts from agriculture to nonagricultural sectors. This overall process can be called "agricultural transformation," and the contribution of each of the factors has been quantified in the existing literature (Timmer 1988).

[2] Datasets are newly compiled by the author, using government statistics. Using the previous versions of these datasets, Kurosaki (1999) and Kurosaki (2002) investigated the performance of agriculture in India and Pakistan for the period c.1900-1995, Kurosaki (2003) quantified the growth impact of crops shifts in West Punjab, Pakistan for a similar period, and Kurosaki (2006) extended the analysis for India and Pakistan using updated estimates until 2004.

[3] For example, Blyn (1966) and Sivasubramonian (1960, 1997);see also references in Kurosaki (1999).

tion. The third section gives an analytical framework to structurally associate changes in aggregate land productivity with inter-crop and inter-district reallocations of land use. The fourth section presents empirical results, which show that crop shifts did contribute to agricultural growth in India and Pakistan. The final section concludes the chapter.

CHANGES IN LAND UTILIZATION IN INDIA AND PAKISTAN

The Indo-Gangetic Plains currently extends into three countries of India, Pakistan, and Bangladesh. India, with an area of about 3 million km^2 of land and a population of 1.1 billion, obtained independence from the British in August 1947. Contemporary Pakistan has a land area of 800,000 km^2 and a population of over 150 million people. It became independent in August 1947, partitioned from India to form a country comprising its west wing (today's Pakistan) and its east wing (today's Bangladesh, which obtained independence from West Pakistan in 1971). Bangladesh has a land area of 140,000 km^2 and a population of 140 million.

In August 1947, the Indian Empire under British rule was partitioned into India and (United) Pakistan. Before 1947, the Empire was subdivided into provinces of British India and a large number of Princely States. The current international borders are different, not only from provincial/state borders, but also from boundaries of districts (the basic administrative unit within a province). The two important provinces of Bengal and Punjab were divided into India and (United) Pakistan with Muslim majority districts belonging to the latter. In the process, several districts in Bengal and Punjab were also divided.

Before 1947, agricultural statistics were collected regularly in all provinces of British India. In contrast, statistical information on the Princely States is limited in coverage and missing for many regions. Because of this reason, the classic and seminal study on agricultural growth in colonial India by Blyn (1966) examined the area known as 'British India', which covers all British provinces except for Burma (Burma Province became a separate colony in 1937). British India below corresponds to the area defined by Blyn (1966).

Table 4.1 Decade-wise land utilization in India and Pakistan, 1901-2002

In million ha.

	Reported area (total of [1]-[5])	Forest [1]	Not available for cultivation [2]	Cultivable waste [3]	Current fallow [4]	Net area sown [5]	Total area cultivated ([4]+[5])
British India							
1901/02	182.01	22.33	33.66	33.73	16.08	76.20	92.28
(% to the total)	(100.0%)	(12.3%)	(18.5%)	(18.5%)	(8.8%)	(41.9%)	(50.7%)
1911/12	205.31	25.04	42.20	35.95	20.11	82.01	102.12
1921/22	205.52	26.78	39.58	36.45	18.88	83.84	102.72
1931/32	206.88	26.86	37.87	38.49	18.13	85.54	103.67
1941/42	207.25	27.67	36.86	37.32	19.08	86.32	105.40
(% to the total)	(100.0%)	(13.3%)	(17.8%)	(18.0%)	(9.2%)	(41.6%)	(50.9%)
Annual growth rate from	0.32%	0.54%	0.23%	0.25%	0.43%	0.31%	0.33%
India							
1951/52	287.83	48.89	50.17	40.40	28.96	119.40	148.36
(% to the total)	(100.0%)	(17.0%)	(17.4%)	(14.0%)	(10.1%)	(41.5%)	(51.5%)
1961/62	305.35	60.84	50.36	36.58	21.23	136.34	157.57
1971/72	304.02	65.41	41.82	34.66	24.53	137.59	162.12
1981/82	304.11	67.35	39.95	31.85	24.17	140.79	164.97
1991/92	304.84	67.98	40.91	29.40	23.83	142.72	166.55
2001/02	305.01	69.51	41.78	27.36	24.95	141.42	166.36
(% to the total)	(100.0%)	(22.8%)	(13.7%)	(9.0%)	(8.2%)	(46.4%)	(54.5%)
Annual growth rate from	0.12%	0.70%	-0.37%	-0.78%	-0.30%	0.34%	0.23%
Pakistan							
1951/52	46.45	1.39	20.75	9.16	3.54	11.61	15.15
(% to the total)	(100.0%)	(3.0%)	(44.7%)	(19.7%)	(7.6%)	(25.0%)	(32.6%)
1961/62	50.99	1.68	18.73	12.46	4.85	13.27	18.12
1971/72	53.55	2.83	20.40	11.11	4.77	14.44	19.21
1981/82	53.92	2.85	19.90	10.86	4.89	15.41	20.30
1991/92	57.61	3.46	24.34	8.85	4.85	16.11	20.96
2001/02	59.28	3.61	24.50	9.13	5.67	16.32	21.99
(% to the total)	(100.0%)	(6.1%)	(41.3%)	(15.4%)	(9.6%)	(27.5%)	(37.1%)
Annual growth rate from	0.49%	1.91%	0.33%	-0.01%	0.94%	0.68%	0.75%

Data sources: British India (excluding Burma):Agricultural Statistics of India, Government of India, various issues; India: Indian Agricultural Statistics, Directorate of Economics and Statistics, Ministry of Agriculture, Government of India, various issues; Pakistan (corresponding to contemporary Pakistan): Economic Survey, Ministry of Finance, Government of Pakistan, various issues.

In 1901/02,[4] out of 182 million hectares of land for which the information was available, 12.3 percent was under forest and 50.7 percent was under cultivation. About 17 percent of the total cultivated land was fallow. In 1941/42, these shares were similar but the absolute acreage of land under forest or land under cultivation increased, at annual growth rates of 0.54 percent and 0.33 percent respectively. Besides the land under forest or under cultivation, there was a huge area that was not available for cultivation or was classified as cultivable waste. Most of these lands were barren, with very limited vegetation.

Table 4.1 also shows decade-wise land utilization in India and Pakistan after independence. Each series shows statistics for a geographic area corresponding to the current international borders of India and Pakistan. In 1951/52, just after the partition, 17.0 percent of the reported land was under forest and 51.5 percent was under cultivation in India, higher than corresponding figures for Pakistan (3.0% and 32.6%). This shows that Pakistan inherited more barren land than India did.

Both in India and in Pakistan, the area under forests and that under cultivation increased substantially throughout the post-independence period. The growth rates were higher in Pakistan than in India: the forest and total cultivated areas increased at annual growth rates of 1.91 percent and 0.75 percent respectively in Pakistan, well above the figures for British India before independence. In India, these growth rates were lower than in Pakistan but were still comparable to the rates recorded before independence. It is worth mentioning that the area not available for cultivation or cultivable waste decreased in India. The post-independence expansion of the cultivated area was even more impressive if we take into account the area sown more than once during the agricultural year. Such changes in cropping patterns are the theme of the rest of this chapter. Regarding cropping patterns and crop output, there are several sources of information covering the Princely States. Utilizing these sources, Kurosaki (2006) compiled an updated version of the country-level dataset compiled for India and Pakistan, covering a period from 1901/02 to 2003/04, and covering the production of principal crops that are important in contemporary India and Pakistan.[5]

---

[4] "1901/02" refers to the agricultural year in India and Pakistan beginning on July 1, 1901, and ending on June 30, 1902. In figures with limited space, it is shown as "1902."

[5] For India, eighteen crops are included: rice, wheat, barley, *jowar* (sorghum), *bajra* (pearl millet), maize, *ragi* (finger millet), *gram* (chickpea), linseed, sesamum, rapeseed & mustard, groundnut, sugarcane, tea, coffee, tobacco, cotton, and jute & mesta. These crops currently account for more than two thirds of the total output

The data compilation procedure for the colonial period is explained by Kurosaki (1999). Data on the areas that are currently in Pakistan and Bangladesh were subtracted from the database compiled by Sivasubramonian (1960, 1997). Information included in the district-level data in *Season and Crop Reports* from Punjab, Sind (or Bombay-Sind), the North-West Frontier Province, and Bengal was utilized in this exercise.

ANALYTICAL FRAMEWORK

To analyze the growth performance of agriculture in India and Pakistan, the gross output values of these crops are aggregated using 1960 prices,[6] and denoted by Q. As measures for partial productivity, Q is divided by L (the official population estimates of India and Pakistan) or by A (the sum of the acreage under the twelve or eighteen crops). As the first step to analyze the changes in agricultural productivity, a time series model for Y_t is estimated as

$$(1) \quad \ln Y_t = a_0 + a_1 t + u_t ,$$

where t is measured in years and u_t is an error term. Equation (1) is estimated for the logarithm of Q, Q/L, and Q/A, by the ordinary least squares (OLS) method. The larger the coefficient estimate for a_1, the higher the growth rate of production or productivity.

Then in the next step, to capture long-term changes in the crop mix, the Herfindahl Index of crop acreage was calculated. Let S_i be the acreage share of crop i in the sum of the principal crops. The Herfindahl Index is defined as

---

value from the crop sector and more than half of the total output from agriculture, and their contribution was higher in the colonial period. For Pakistan, twelve major crops are covered: rice, wheat, barley, *jowar*, *bajra*, maize, *gram*, rapeseed & mustard, sesamum, sugarcane, tobacco, and cotton. These crops currently account for about 70 percent of value-added of all crops in Pakistan and about 40% of value-added of agriculture, and their share, similarly, was higher in the colonial period.

[6] Ideally, the sum of the value-added evaluated at current prices and then deflated using a price index would be a better measure, but the sum of gross output values at constant prices is used as a proxy due to the absence of reliable data on input prices and quantities before independence. The results reported in this chapter are insensitive to the choice of base year (1938/39 and 1980/81).

$$(2) \quad H = \sum_i S_i^2 ,$$

which can be intuitively understood as the probability of hitting the same crop when two points are randomly chosen from all the land under consideration. Therefore, a higher H implies a greater concentration of acreage into a smaller number of crops.

The traditional approach in analyzing agricultural productivity is through growth accounting, estimating the total factor productivity (TFP) as a residual after controlling for factor inputs (Timmer 1988). As a complement to the TFP approach, Kurosaki (2003) proposed a methodology to focus on the role of resource reallocation within agriculture --- across crops and across regions. Unlike in manufacturing industries, the spatial allocation of land is critically important in agriculture due to high transaction costs including transportation costs (Takayama and Judge 1971; Baulch 1997). Because of this, farmers may optimally choose a crop mix that *does not* maximize expected profits evaluated at market prices but *does* maximize expected profits evaluated at farm-gate prices after adjusting for transaction costs (Omamo 1998a; 1998b). Subjective equilibrium models for agricultural households provide other reasons for the divergence of decision prices by farmers from market prices. In the absence of labor markets, households need to be self-sufficient in farm labor (de Janvry et al. 1991), and if insurance markets are incomplete, farmers may consider production and consumption risk or the domestic needs of their families (Kurosaki and Fafchamps 2002). In these cases, their production choices can be expressed as a subjective equilibrium evaluated at household-level shadow prices.

During the initial phase of agricultural transformation, therefore, it is likely that the extent of diversification will be similar at the country level and the more micro levels because, given the lack of well-developed agricultural produce markets, farmers have to grow the crops they want to consume themselves (Timmer 1997). As rural markets develop, however, the discrepancy between the market price of a commodity and the decision price at the farm level is reduced. In other words, the development of rural markets is a process which allows farmers to adopt production choices that reflect their comparative advantages more closely, and thus contributes to productivity improvement at the aggregate level evaluated at common, market prices. Therefore, the effect of crop shifts on productivity growth is a useful indicator of market development in developing countries.

To quantify this effect, changes in aggregate land productivity can be decomposed into crop yield effects, static inter-crop shift effects, and dynamic inter-crop shift effects (Kurosaki 2003). Let Y_t denote per-acre output in year t. Its growth rate from period 0 to period t can be decomposed as

$$(3)(Y_t - Y_0)/Y_0 = [\sum_i S_{i0}(Y_{it} - Y_{i0}) + \sum_i (S_{it} - S_{i0})Y_{i0} + \sum_i (S_{it} - S_{i0})(Y_{it} - Y_{i0})]/Y_0$$

where the subscript i denotes each crop so that Y_{it} stands for per-acre output of crop i in year t. The first term of equation (3) captures the contribution from the productivity growth of individual crops. The second term shows 'static' crop shift effects, as it becomes more positive when the area under crops whose yields were initially high increases in relative terms. The third term shows 'dynamic' crop shift effects, as it becomes more positive when the area under dynamic crops (i.e., crops whose yields are improving) increases relative to the area under non-dynamic crops.

For each crop, another aspect of land-use changes can be investigated, focusing on the effect of *inter-spatial* crop shifts on land productivity. Kurosaki (2003) thus proposed a further decomposition of the crop yield effect for crop i in equation (3) as

$$(4) \quad Y_{it} - Y_{i0} = \sum_h S_{hi0}(Y_{hit} - Y_{hi0}) + \sum_h (S_{hit} - S_{hi0})Y_{hi0} + \sum_h (S_{hit} - S_{hi0})(Y_{hit} - Y_{hi0})$$

where S_{hit} is the share of district h in the cultivated area of crop i in year t. The three terms on the right hand side of equation (4) are interpreted similarly to the terms in equation (3): the first term shows the effects of the average crop yields in the district ('District crop yield effects'), the second term indicates '*Inter-district* crop shift effects (static)' and the third term shows '*Inter-district* crop shift effects (dynamic)'. An important aspect of equation (4) is that the effects of factor reallocation over space are incorporated explicitly into the decomposition. In other words, so-called 'yield effects' in the existing literature based on macro data are often a mixture of pure yield effects (e.g., due to shifts in TFP in producing individual crops) and spatial crop shift effects.

EMPIRICAL RESULTS: AGRICULTURAL GROWTH AND CROP SHIFTS IN INDIA AND PAKISTAN

Productivity

Growth performance in Indian agriculture is plotted in Figure 4.1. Table 4.2 shows estimation results for equation (1). The total output value (Q)

grew very little in the period before independence in 1947 and then grew steadily afterward. The growth rate in the 1990s was 1.7 percent, a rate lower than the post-independence average of 2.7 percent. The column 'Co-eff.var' shows the level of variability of the output around the fitted values. The 1990s were associated with less variability. In the 1990s, output per capita (*Q*/*L*) did not grow while output per acre (*Q*/*A*) grew at a rate similar to that of *Q*. In other words, the source of growth in the 1990s in India was exclusively an improvement in aggregate land productivity.

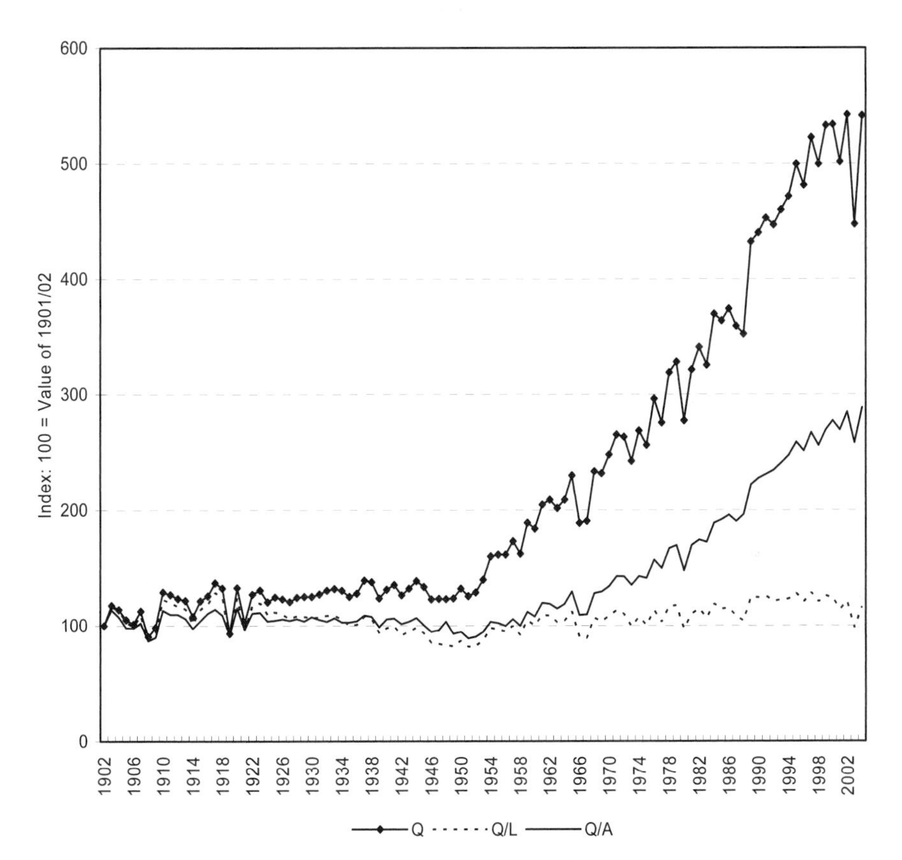

Fig 4.1. Agricultural output in India, 1901/02-2003/04

Similar changes were observed in Pakistan but with higher growth rates throughout the period (Figure 4.2, Table 4.2). The growth rate of *Q* before independence in 1947 was 1.3 percent, smaller than that after independence (3.5%) but still statistically significant. During the post-independence period, the growth rate declined to 2.3 percent in the 1990s, but still was higher than that of India. Unlike in India, the variability of *Q* was similar to the previous periods and the source of growth was not only an improvement in aggregate land productivity but also an expansion of total farmed acreage. The growth rate of output per acre (Q/A) in the 1990s was positive and significant but smaller than that of Q.

Table 4.2. Growth performance of agriculture in India and Pakistan, 1901-2004

	Q (Total output value)		Q/L (Output per capita)		Q/A (Output per acre)	
	Growth rate	Coeff.var.	Growth rate	Coeff.var.	Growth rate	Coeff.var.
India						
1901/02 - 1910/11	1.04%	11.8%	0.39%	11.8%	-0.26%	9.8%
1911/12 - 1920/21	-0.88%	13.0%	-0.97%	13.0%	-0.40%	8.0%
1921/22 - 1930/31	-0.08%	2.6%	-1.09% ***	2.6%	-0.41%	2.4%
1931/32 - 1940/41	0.24%	4.0%	-1.16% **	4.0%	0.10%	3.0%
1941/42 - 1950/51	-0.53%	4.2%	-1.78% ***	4.2%	-1.45% **	4.0%
1951/52 - 1960/61	4.24% ***	5.1%	2.28% ***	5.1%	2.34% ***	4.2%
1961/62 - 1970/71	2.53% **	8.8%	0.32%	8.8%	1.89% **	7.2%
1971/72 - 1980/81	2.62% **	7.1%	0.41%	7.1%	2.12% ***	5.6%
1981/82 - 1990/91	3.21% ***	6.2%	1.07%	6.2%	3.23% ***	3.7%
1991/92 - 2000/01	1.68% ***	3.5%	-0.27%	3.5%	1.62% ***	2.4%
1901/02 - 1946/47	0.48% ***	8.5%	-0.34% ***	9.6%	-0.01%	6.3%
1947/48 - 2003/04	2.72% ***	7.5%	0.60% ***	7.6%	2.19% ***	6.3%
Pakistan						
1901/02 - 1910/11	4.32% **	15.1%	2.75%	15.1%	0.99%	10.9%
1911/12 - 1920/21	-0.33%	14.6%	-1.18%	14.6%	-0.19%	6.6%
1921/22 - 1930/31	-0.64%	10.3%	-1.73%	10.3%	-1.15%	7.4%
1931/32 - 1940/41	2.81% ***	5.8%	0.97%	5.8%	1.86% **	5.1%
1941/42 - 1950/51	0.05%	6.7%	-2.92% ***	6.7%	-0.19%	3.4%
1951/52 - 1960/61	3.44% ***	5.2%	1.00%	5.2%	1.66% ***	3.3%
1961/62 - 1970/71	5.85% ***	5.2%	2.99% ***	5.2%	3.93% ***	4.5%
1971/72 - 1980/81	3.24% ***	3.7%	0.09%	3.7%	1.75% ***	3.3%
1981/82 - 1990/91	3.50% ***	5.2%	0.85%	5.2%	2.65% ***	5.3%
1991/92 - 2000/01	2.30% ***	5.3%	-0.35%	5.3%	1.61% **	5.5%
1901/02 - 1946/47	1.30% ***	12.8%	-0.03%	11.9%	0.38% ***	8.6%
1947/48 - 2003/04	3.48% ***	8.2%	0.68% ***	7.7%	2.30% ***	6.4%

Source: Kurosaki (1999) and updated estimates by the author.
Note: 'Growth rate' provides a parameter estimate for the slope of the log of Q (or Q/L or Q/A) on a time trend, estimated by OLS (see equation (1)). The parameter estimate is statistically significant at 1% ***, 5% **, or 10% * (two sided t-test). 'Coeff.var' shows the coefficient of variation approximated by the standard error of the OLS regression

In both countries in the post-independence period, the growth rate of output per capita (Q/L) was much smaller than that of output per acre (Q/A). Since Q/L is more closely related with the number of the absolutely poor than Q/A, this is a worrying situation. Using the identity $Q/L = (Q/A)*(A/L)$, the low growth rate of Q/L can be interpreted as the combination of a positive but not sufficiently high growth rate of Q/A and a negative growth rate of A/L. Farmland expansion was lagging behind the population growth in India and Pakistan.

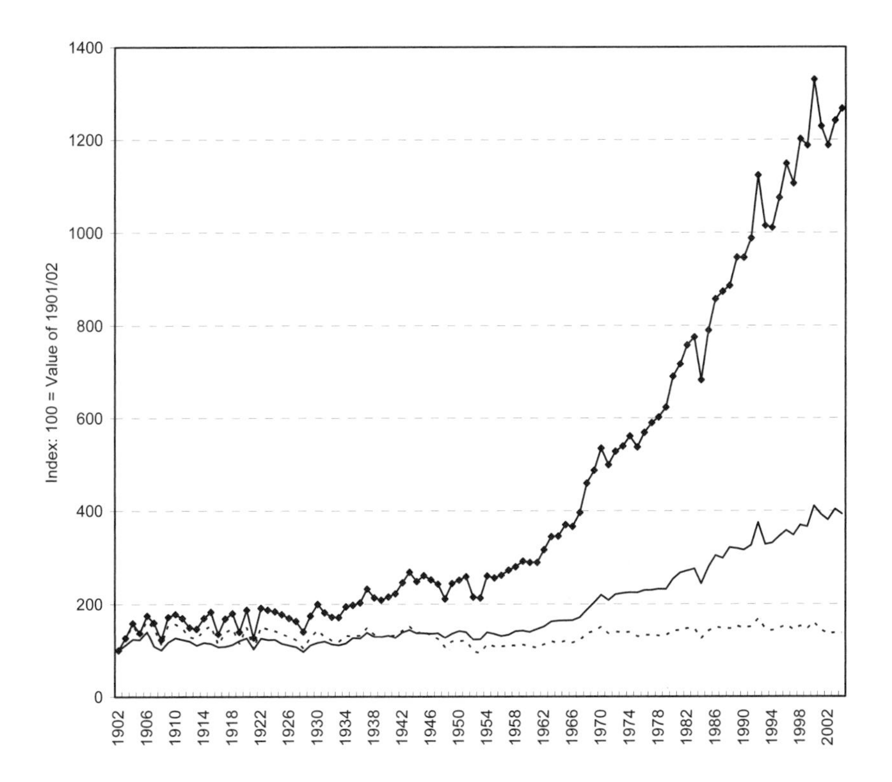

Fig. 4.2. Agricultural output in Pakistan, 1901/02-2003/04

Changes in Crop Mix

As shown in Figure 4.3, the post-independence period was associated with increasing H over time in India. In the 1990s, this trend was accelerated. The level of concentration seems to have reached a plateau in the early 2000s.

To investigate whether these changes in crop mix were consistent with those indicated by comparative advantage and market development, decomposition (3) was implemented (Table 4.3). Throughout the post-independence period, there were substantial contributions from both static and dynamic crop shift effects. More interestingly, during the 1990s, the growth due to improvements in crop yields was reduced compared to the 1980s while the growth due to static crop shifts was higher. As a result, the relative contribution of static shift effects was as high as 39 percent in the 1990s. This is the highest figure for all the post-independence decades. Therefore, it can be concluded that the changes in crop mix in the 1990s (the decade of economic liberalization in India) were indeed consistent with the comparative advantages of Indian agriculture so that the aggregate land productivity was improved.

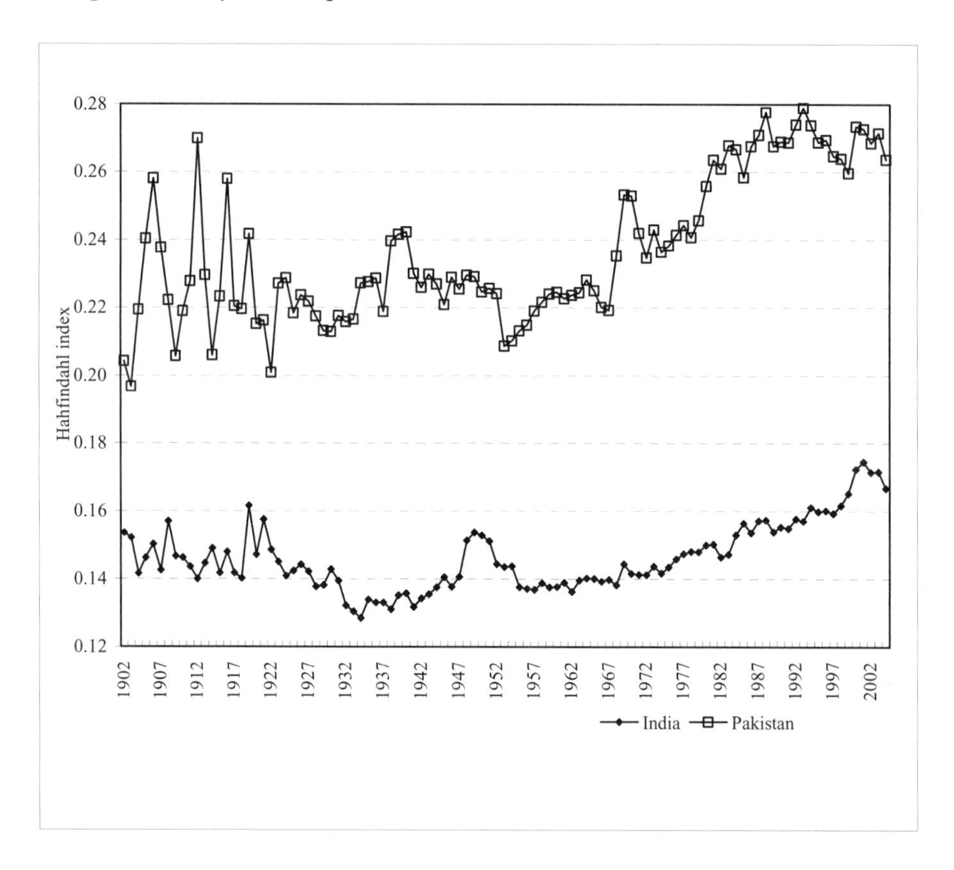

Fig. 4.3. Crop concentration in India and Pakistan, 1901/02-2003/04

In contrast, the crop concentration index in Pakistan did not accelerate in the 1990s (Figure 4.3). Rather it remained at the high level that had already been reached during the late 1980s or early 1990s. This seems to indicate that shifts in acreage toward crops with comparative advantages occurred earlier in Pakistan than in India, possibly reflecting Pakistan's attempt to liberalize agricultural marketing during the early 1980s.

Table 4.3. Contribution of crops shifts to land productivity growth in India and Pakistan

	Annural growth rates of land productivity (%)				Contribution share (%)		
	Pure yield effects	Static shift effects	Dynamic shift effects	Total	Pure yield effects	Static shift effects	Dynamic shift effects
India							
1901/02 - 1911/12	0.90	0.00	-0.04	0.85	105.1	-0.1	-5.1
1911/12 - 1921/22	-0.35	-0.07	0.26	-0.17	209.5	43.8	-153.3
1921/22 - 1931/32	-0.34	0.14	0.05	-0.14	234.6	-97.2	-37.4
1931/32 - 1941/42	-0.36	0.30	-0.06	-0.12	290.9	-239.7	48.8
1941/42 - 1951/52	-1.48	0.30	-0.01	-1.20	124.0	-25.1	1.1
1951/52 - 1961/62	2.76	0.14	-0.01	2.89	95.3	5.0	-0.3
1961/62 - 1971/72	1.55	0.15	0.20	1.90	81.6	8.0	10.3
1971/72 - 1981/82	1.83	0.35	0.09	2.28	80.6	15.4	4.0
1981/82 - 1991/92	3.10	0.44	0.14	3.68	84.2	12.1	3.7
1991/92 - 2001/02	0.87	0.59	0.06	1.51	57.5	38.7	3.7
1901/02 - 1947/48	-0.24	-0.05	0.23	-0.06	423.5	82.2	-405.7
1947/48 - 2003/04	2.79	0.21	0.63	3.63	76.8	5.8	17.4
Pakistan							
1901/02 - 1911/12	1.84	-0.19	0.09	1.74	105.4	-10.8	5.4
1911/12 - 1921/22	0.06	-0.05	0.02	0.02	254.3	-226.2	71.9
1921/22 - 1931/32	-0.35	0.03	0.02	-0.31	113.4	-8.5	-4.9
1931/32 - 1941/42	1.51	0.16	0.32	1.99	75.8	8.2	16.0
1941/42 - 1951/52	-0.80	0.18	0.03	-0.58	136.6	-30.6	-6.0
1951/52 - 1961/62	1.03	0.85	0.03	1.92	53.8	44.4	1.8
1961/62 - 1971/72	3.37	0.52	0.28	4.16	80.8	12.5	6.7
1971/72 - 1981/82	1.72	0.63	0.13	2.49	69.2	25.4	5.4
1981/82 - 1991/92	2.35	0.07	0.21	2.63	89.3	2.5	8.2
1991/92 - 2001/02	1.19	0.23	0.02	1.43	82.8	16.1	1.1
1901/02 - 1947/48	0.55	-0.03	0.22	0.74	74.5	-4.0	29.6
1947/48 - 2003/04	2.50	0.48	0.61	3.59	69.7	13.3	17.0

Source: Kurosaki (2002) and update estimates by the author.
Note: Annual growth rates were estimated using the method explained in the text (see equation (3). Since both the estimate model and the data treatment for smoothing are different, the total growth rates of land productivity in this table are slightly different from those shown in Table 4.2.

The results of the decomposition for Pakistan in Table 4.3 show that there were substantial contributions by both static and dynamic crop shift effects, and that their shares were larger than in India throughout the post-independence period. This indicates that crop shift effects are more important in a smaller economy. During the 1990s, the growth due to improvements in crop yields declined substantially while the growth due to static crop shifts recovered. As a result, the relative contribution of static shift effects was 16 percent in the 1990s, a level higher than the post-independence average (13%). Here we find a similarity between India and Pakistan: in both economies, the crop shifts were an important source of land productivity growth in the post-independence period, and especially in the 1990s.

Analysis at the District Level

The contribution of crop shifts to productivity growth can be analyzed in more detail using more geographically disaggregated data. For this purpose, Kurosaki (2003) compiled district-level panel data on agricultural production for fifteen districts in West Punjab, Pakistan.[7] First, the Herfindahl Index (H) for crop concentration was calculated for each district. Figure 4 plots the median, upper quartile and lower quartile of H in each year for the 15 districts, together with the same index at the aggregate level for comparison. The exact districts ranked in each position are not the same throughout the period, but there is no major churning among districts.

The figure shows first that the crop mix is more diversified at the more aggregate level than at the district level throughout the period. Second, the concentration indices show an increase in the latter half of the study period. Thus, the acceleration of aggregate land productivity growth in Figure 2 since the mid-1950s was associated with an increasing concentration index in each district. Third, the difference between aggregate and district-median indices widens in the latter half. Fourth, as shown by the upper and lower quartile plots, the dynamic paths differ widely from district to district.

Based on these records, Kurosaki (2003) hypothesized that the shift of cultivated areas from less lucrative to more lucrative crops and from less productive to more productive districts was an important source of agricultural growth in West Punjab. The decomposition results according to equation (3) supported the first part of this hypothesis.[8]

Regarding the second hypothesis, Table 4.4 shows the decomposition results according to equation (4) for four major crops in West Punjab. For

---

[7] The 15 districts of West Punjab accounted for approximately 55percent of farm land in Pakistan and 65 percent of agricultural gross output in 1981.

[8] See Table 4.2 in Kurosaki (2003), which gives qualitatively the same results as Table 4.3 in this chapter for Pakistan.

wheat and sugarcane, aggregate yield improved mainly through improvements in crop yields at the district level. In contrast, inter-district crops shift effects were the major source of improvement in the aggregate yields for rice and cotton. In the first period, when aggregate rice yields grew at 0.27 percent (statistically significant at the 1% level), the major source of yield growth was crop shifts across districts --- more than 60 percent of yield growth was attributable to static and dynamic crop shift effects in each case

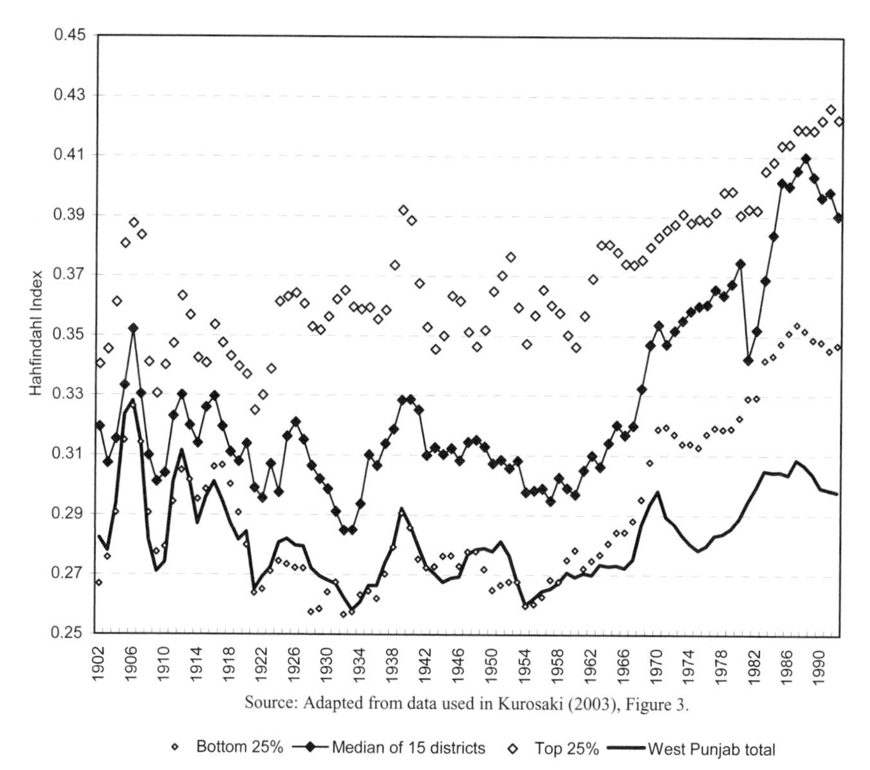

Source: Adapted from data used in Kurosaki (2003), Figure 3.

Fig. 4.4. District-level crop concentration in West Punjab, Pakistan, 1901/02-1991/92

In the case of cotton, dynamic crop shift effects were important in both periods. They explain more than one fourth of the aggregate yield growth for cotton. On the other hand, static crop shift effects were nil. This implies that the improvement of cotton yields in West Punjab was facilitated by the shifting of cotton areas from districts with stagnating or decreasing productivity to districts with increasing productivity. The expansion of cotton production for both domestic and foreign markets was the most important development in Punjab's agriculture during the colonial period. Still today, Pakistan's economy remains heavily dependent on cotton production in

West Punjab. The analysis here shows that the land productivity of cotton improved not only through improvements of crop yields at the district level but also through a reallocation of cultivated land to districts experiencing rapid improvements in cotton yields.

Table 4.4. Contribution of inter-district crop shifts to crop yield growth in West Punjab, Pakistan

	Average Annual Growth Rates (%)			
	District crop yield effects	Inter-district crop shift effects (static)	Inter-district crop shift effects (dynamic)	Total
A. Wheat				
1911/12 - 1951/52	-0.03	0.02	0.06	0.04
	(-76.8)	(40.2)	(136.7)	(100.0)
1951/52 - 1991/92	2.69	0.08	0.10	2.86
	(93.9)	(2.8)	(3.3)	(100.0)
B. Rice				
1911/12 - 1951/52	-0.08	0.18	0.17	0.27
	(-27.4)	(64.0)	(63.4)	(100.0)
1951/52 - 1991/92	0.70	-0.03	-0.04	0.63
	(110.2)	(-4.2)	(-6.0)	(100.0)
C. Sugarcane				
1911/12 - 1951/52	2.44	0.13	-0.04	2.53
	(96.4)	(5.0)	(-1.5)	(100.0)
1951/52 - 1991/92	0.52	0.01	0.00	0.53
	(98.2)	(2.0)	(-0.2)	(100.0)
D. Cotton				
1911/12 - 1951/52	1.26	-0.04	0.44	1.66
	(75.9)	(-2.5)	(26.7)	(100.0)
1951/52 - 1991/92	4.22	0.01	1.72	5.96
	(70.8)	(0.2)	(28.9)	(100.0)

Source: Kurosaki (2003), Table 3;
Notes: Numbers in parenthesis show relative contribution to growth (%). Kurosaki (2003) estimated the timing of a breakdate using a time-series model and chose the year 1951/52 as the breakdate.

The evidence presented in Table 4.4 shows the importance of inter-district crop shifts for cash crops such as rice and cotton in facilitating yield growth at the aggregate level. This suggests that the historical change in West Punjab is consistent with crop shifts reflecting static and dynamic comparative advantage, a finding consistent with the country-level analysis for India and Pakistan

CONCLUSION

Based on a production dataset from India and Pakistan for the period 1901-2004, this chapter described changes in land use, associated the changes with long-term agricultural performance, and showed the importance of crop shifts in enhancing agricultural productivity. The empirical results showed a discontinuity between the pre- and the post-independence periods, both in India and in Pakistan. Total output growth rates rose from zero or very low figures to significantly positive levels, which were sustained throughout the post-independence period; the crop mix changed with increasing concentration beginning in the mid 1950s. This chapter quantified the effects of *inter-crop* and *inter-district* crop shifts, a previously unnoticed source of productivity growth, on land productivity. It was found that the crop shifts did contribute to the productivity growth, especially during the periods with limited technological breakthroughs.

Underlying these effects were the responses of farmers to changes in market conditions and agricultural policies. Agriculture in these regions experienced a consistent concentration of crops after the mid 1950s, when agricultural transformation in terms of output per agricultural worker was proceeding. These trends continued until the early 1990s in Pakistan and until the early 2000s in India. The performance in the latest periods suggests that agriculture in these regions seems to have entered a new phase of diversified production and consumption at the country level (Timmer 1997). Whether the trends will reverse remains to be investigated when more recent data become available, it is of interest to note that the concentration trended to reach a plateau earlier in Pakistan than in India. This contrast can be attributed to the earlier adoption of economic liberalization policies in Pakistan in the early 1980s. Pakistani farmers were exposed to international prices relatively earlier than their Indian counterparts.

In both countries in the post-independence period, however, the growth rate of aggregate land productivity was not high enough to cancel the negative growth rate of land availability per capita. The net result was that the growth rate of agricultural output per capita was much smaller than that of output per acre, resulting in a slow pace of poverty reduction in these two countries. The crop shift effects identified in this chapter was not sufficiently strong in this sense. Reducing population growth rates and absorbing more labor force outside agriculture are required to make the growth rate of per-capita agricultural output comparable to that of per-acre agricultural output.

Although this chapter showed the importance of crop shifts in improving aggregate land productivity, the overall impact is underestimated, because only major crops were covered. Incorporating non-traditional crops into the framework of this chapter would be highly desirable. To quantify the structural determinants of these changes and their net effects on the welfare of rural population, further research is needed, such as an analysis of production costs, investigation of minor crops and livestock activities, etc. These are left for future study.

REFERENCES

Baulch B (1997) Transfer Costs, Spatial Arbitrage, and Testing Food Market Integration. American Journal of Agricultural Economics 79:477-487

Blyn G (1966) Agricultural Trends in India, 1891-1947: Output, Availability, and Productivity. University of Pennsylvania Press, Philadelphia

de Janvry A, Fafchamps M, Sadoulet E (1991) Peasant Household Behavior with Missing Markets: Some Paradoxes Explained. Economic Journal 101:1400-17

Kurosaki T (1999) Agriculture in India and Pakistan, 1900-95: Productivity and Crop Mix. Economic and Political Weekly 34:A160-A168

Kurosaki T (2002) Agriculture in India and Pakistan, 1900-95: A Further Note. Economic and Political Weekly 37:3149-3152

Kurosaki T (2003) Specialization and Diversification in Agricultural Transformation: The Case of West Punjab, 1903-1992. American Journal of Agricultural Economics 85:372-386

Kurosaki T (2006) Long-term Agricultural Growth and Crop Shifts in India and Pakistan. Journal of International Economic Studies 20:19-35

Kurosaki T, Fafchamps M, (2002) Insurance Market Efficiency and Crop Choices in Pakistan. Journal of Development Economics 67:419-453

Omamo SW (1998a) Transport Costs and Smallholder Cropping Choices: An Application to Siaya District, Kenya. American Journal of Agricultural Economics 80:116-123

Omamo SW (1998b) Farm-to-Market Transaction Costs and Specialization in Small-Scale Agriculture: Explorations with a Non-separable Household Model. Journal of Development Studies 35:152-163

Sivasubramonian S (1960) Estimates of Gross Value of Output of Agriculture for Undivided India 1900-01 to 1946-47. In: Rao VKRV, Sen SR, Divatia MV, Uma Datta (eds) Papers on National Income and Allied Topics, Asia Publishing House, New York, Vol. I, pp 231-251

Sivasubramonian S (1997) Revised Estimates of the National Income of India, 1900-01 to 1946-47. Indian Economic and Social History Review 34:113-168

Takayama A, Judge G (1971) Spatial and Temporal Price and Allocation Models. North Holland, Amsterdam

Timmer CP (1988) The Agricultural Transformation. In: Chenery H, Srinivasan TN (eds) Handbook of Development Economics, Elsevier Science, Amsterdam, Vol I, pp 275-331

Timmer CP (1997) Farmers and Markets: The Political Economy of New Paradigms. American Journal of Agricultural Economics 79:621-627

World Bank (2001) World Development Report 2000/2001: Attacking Poverty. Oxford University Press, Oxford

Chapter 5 - Agricultural Intensification on Brazil's Amazonian Soybean Frontier

Wendy Jepson[1], J. Christopher Brown[2], and Matthew Koeppe[3]

[1] *Department of Geography, Texas A&M University, College Station, TX 77843-3147, USA*
[2] *Department of Geography and Environmental Studies Program, University of Kansas, Lawrence, KS 66045-7613, USA*
[3] *Department of Geography, The George Washington University, Washington, DC 20052 USA*

ABSTRACT

Mechanized agriculture in Brazil's Amazon Basin predominantly occurs in the states of Mato Grosso and Rondônia, with a production trend moving from south to north. We focus our attention on the land-change imprint of capital-intensive, mechanized agricultural operations, particularly land-use intensification on the soybean frontier region in the municipality of Vilhena, Rondônia. Assessing the land-change imprint of modern agriculture in the Brazilian Amazon, however, requires a step-by-step approach that separates land-change analysis of land-cover classes (e.g., forest to agriculture) from changes within classes (e.g., pasture to agriculture; rice production to soybean production). With the moderate spatial resolution, high temporal resolution satellite remote sensing, it is possible to obtain data on land modification processes related to agricultural intensification. Previous research has shown that cropping frequency, as a surrogate for agricultural intensity, can be mapped on an annual basis using the Moderate Resolution Imaging Spectroradiometer (MODIS). This research presents research that links land tenancy of mechanized farming to cropping frequency data extracted from MODIS 250 m imagery from 2001 to 2005.

INTRODUCTION

Mechanized agriculture in Brazil's Amazon Basin predominantly occurs in the states of Mato Grosso and Rondônia, with a production trend moving from south to north (Simon and Garagorry 2005). The path of development follows private colonization and agronomic technical change (Brown et al. 2005b; Jepson 2006a), processes that are distinct from the typical economic geographies of state-led colonist settlement and cattle ranching characteristic of development in the Brazilian Amazon during the 1970s and 1980s (Lisansky 1988; Hecht and Corburn 1990; Schmink and Wood 1992). Indeed, mechanized cultivation in the region emerged from two decades of mechanization and technological advances that were first introduced to regional agriculture in the *Cerrado*, the neotropical savanna of Brazil's Center-West (Alho and Martins 1996; Smith et al. 1998; Smith et al. 1999; Warnken 1999; Jepson 2006b). In this chapter, we focus our attention on the land-change imprint of capital-intensive agricultural operations, particularly land-use intensification on the soybean frontier region of southern Rondônia.

Understanding why and how societies intensify agriculture and how intensification affects the environment have shaped a large corpus of research over the past fifty years (e.g., Boserup 1965; Turner and Brush 1987). Agricultural intensification aims to raise productivity, measured as increased crop yields per unit area and unit time. Population increases and changes in consumption patterns have been, and remain, driving forces of agricultural intensification. In response to these pressures and changing economies, farmers, from small-scale cultivators to the largest commercial agro-businessmen, employ diverse strategies. Farmers change inputs (labor and fertilizers), alter fallow cycles, increase cropping intensity, or add new agronomic technologies to achieve increased productivity.

Agricultural intensification and its impact on land-cover patterns and processes are central to land-change science (Lambin et al. 2000; Rindfuss et al. 2004; Keys and McConnell 2005). While changes in farmer strategies driving intensification do not have single-cause explanations, there are two general patterns of intensification. Brown et al. (2007) show how remote sensing can be used to study intensification processes in the mechanized farming areas of the western Amazon. They conceive of agricultural intensification occurring in two separate areas in order to better understand the relationship between forest conversion and intensification. Vertical intensification is an increase in total production of areas already open to agriculture (land-cover modification). Horizontal intensification is an increase in total production on recently deforested land, where agricultural activity has directly replaced forests (land conversion). Currently debate over modern agriculture in the Amazon focuses on horizontal intensification, or the direct replacement of forests with fields. We contend that an understanding of both forms of intensification better illustrate the land-cover change imprint of capital-intensive agriculture in the tropics.

Assessing the land-change imprint of modern agriculture in the Brazilian Amazon, however, requires a step-by-step approach that separates land-change analysis of land-cover classes (e.g., forest to agriculture) from changes within classes (e.g., pasture to agriculture; rice production to soybean production). The methods to measure and map land change between classes (land conversion) are well known; measuring more subtle changes that affect the character of land cover without changing overall classification (land modification) have been neglected (Lambin et al. 2003: 214). The objective of this chapter is to move a step toward the goal of assessing the land-change imprint of capital-intensive agriculture by providing preliminary methods to measure land modification. In particular, the chapter focuses on changes cropping frequency as a proxy of agricultural intensification and links these remote-sensing data with farmer-based field data. We illustrate that moderate spatial resolution satellite data (MODIS 250 m resolution products) can be linked to socially relevant variables at the field level, similar to the well-developed approaches of linking high spatial resolution datasets (e.g., SPOT-HRV; Landsat series sensors) to social variables (Liverman et al. 1999). As an example of the linkage, we examine the relationship between farmers' land-tenure arrangements, which frame land-use decisions, and cropping frequency, one measure of agricultural intensification relevant to global land-change studies.

We begin with a brief overview of soybean production and land change in Rondônia. We outline the importance of land tenancy in land-use decision making and provide qualitative data on various land-tenure arrangements in the region. Following this discussion we explore cropping frequency as it relates to land tenure (owner-operated versus tenant-operated). Preliminary results suggest that there are apparent trends that illustrate an association between particular cropping frequencies and land tenure. We conclude with a discussion of the method and approach to the assessment of modern agricultural in the tropics and outline future directions for research.

SOYBEANS IN SOUTHERN RONDÔNIA

Southern Rondônia, previously a cattle ranching frontier at the transition between savanna and humid tropical forests (Fearnside 1989; Mahar 1989; Millikan 1992; Browder 1994), is becoming one of Amazonia's emerging soybean frontiers (Figure 5.1). The region has attracted both agri-business investments and the attention of scientists concerned about the threat soybean production poses to tropical forests (Fearnside 2001; Kaimowitz and Smith 2001; Hecht 2005; Howden 2006; Morton et al 2006; Greenpeace International 2006; Steward 2007). In addition to new agricultural storage and purchasing facilities and the development of new tropical soybean varieties, the construction of a grain shipping port on the Rio Madeira in Porto Velho and shipping facilities near Manaus have greatly reduced overland and river transport costs, making profitable the export of soybeans produced in southern Rondônia.

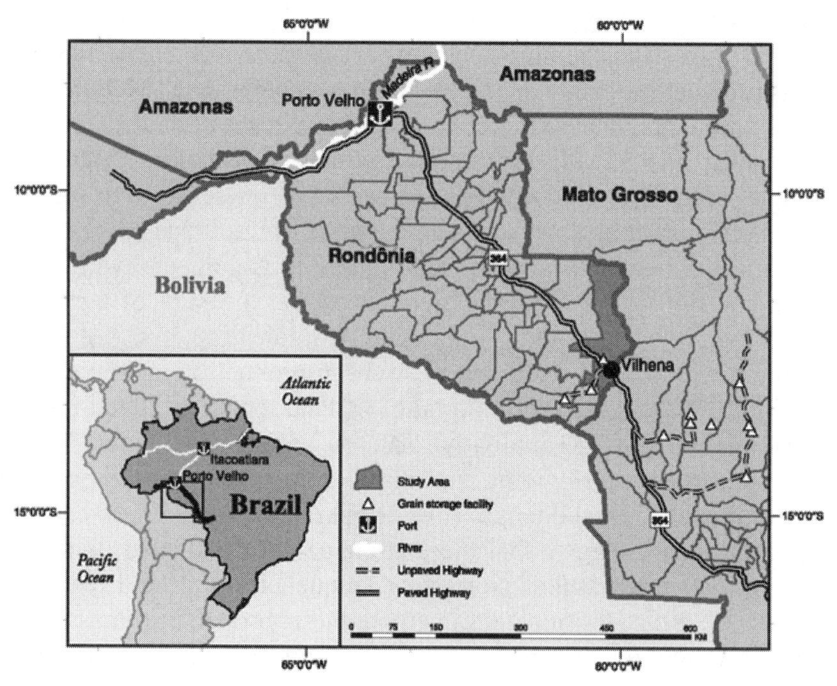

Fig. 5.1. Study area: Vilhena, Rondônia

Land-use and land-cover change

Between 1997 and 2005 soybean production in Rondônia grew considerably, from 656 hectares to 75,275 hectares (IBGE 2005), with new commercial production operations opening every year. A majority of the growth can be attributed to increases in land area dedicated to soybean production in Vilhena, which increased production from 466 hectares in 1997 to 38,000 hectares in 2005 (Table 5.1). However, agricultural census data indicate that mechanized soybean production in the state has been moving west and south, underscored by the fact that Vilhena's contribution to overall state soybean production has decreased, from 72 percent in 2000 to only 50 percent in 2005, as new areas are opening up to mechanized agriculture in adjacent municipalities (IBGE 2005). Colorado d'Oeste, Cerejeiras, and Chupinguaia have seen significant growth. In 2000, these municipalities cultivated 1,100 hectares, 3,400 hectares and 300 hectares, respectively (IBGE 2005). By 2005 soybean production areas increased to 1,500 hectares, 8,000 hectares and 6,500 hectares, respectively (Table 5.1). Of particular note is the expansion of soybean production in Corumbiara, a municipality southwest of Vilhena. In 2000 farmers in Corumbiara did not cultivate soybeans; however, by 2005, the municipality dedicated 12,000 hectares, Rondônia's second-largest area, to its production.

As mechanized farming increased, land covers changed, albeit in surprising trajectories. Brown et al. (2005) use two high-spatial resolution satellite images (1996 and 2000) and an unsupervised, post-classification change detection to measure deforestation due to expansion of mechanized production. The results indicate that primary forests are less likely to be subject to deforestation caused by mechanized agriculture, though they are certainly not immune to it. The study found that cropland increased 12.5 percent (7,036 hectares) between 1996 and 2001. Deforestation represented 43 percent of this land change (2,991 hectares) while conversion of pasture and successional vegetation to mechanized production represented 57 percent (or 4,045 hectares) of change. Deforestation does not account for the total area of soybean production recorded in the official agricultural census for the municipality (Table 5.1). Moreover, Brown and colleagues highlight that deforestation identified in this study is characterized as edge and corridor clearing, i.e. straightening and marginally expanding already existing fields, rather than clear-cutting new swaths of forest for soybean fields. While deforestation due to soybean production remains at the center of environmentalist concerns, these research results point to another process of land change that may be of importance to older Amazonian frontiers: agricultural intensification. Rather than direct conversion from natural to human-influenced land covers, the conversion of successional vegetation and pastures points to land-use modification as an equally important land-change transition in older colonist and cattle ranching regions.

Table 5.1. Area of Soybean Production (Hectares)

	1996	1997	1998	1999	2000	2001	2002	2003	2004	2005
Rondônia	576	656	7,892	7,800	11,800	21,871	28,914	41,600	56,443	75,275
Municipality										
Vilhena	480	466	5,000	5,000	8,500	16,700	22,800	29,000	32,000	38,000
Cabixi	0	0	0	250	200	486	600	1,500	1,500	5,370
Cerejeiras	0	0	2,800	2,300	2,700	3,353	3400	4,516	7,184	8,000
Colorado d'Oeste	1	2	5	250	400	1,027	1,100	1,100	1,400	1,500
Corumbiara	0	0	0	0	0	5	194	600	6,000	12,000
Chupinguaia	0	86	86	0	0	300	300	2,300	5,500	6,500
TOTAL	481	554	7,891	7,800	11,800	21,871	28,394	39,016	53,584	71,370

Source: IBGE 2005

Rondônia Soybean Farmers

Rondonian soybean farmers and regional agronomists were interviewed by the authors between 2003 and 2005. Our understanding of the production system, land tenancy, and the soybean economy is based on eight months of field work in 2004 (conducted by Koeppe) that included repeated semi-structured and informal interviews of farmers, officials of several government agencies[1], input distributors, and three commodity buyers (Maggi, Cargill, and an independent rice buyer). In 2005 Jepson and Brown conducted follow-up field interviews and a survey of 13 farmers or farm managers (from a total population of 32 farm operations) responsible for 17,312 hectares or 45 percent of Vilhena's soybean land in 2005.

Mechanized agriculture production is highly varied, even within the small population of farmers in Vilhena. Soybean producers come from diverse socio-economic backgrounds, levels of capital, and scales of operation. The largest farm operation cultivated 4,700 hectares and the smallest operation worked on 278 hectares with the average land holding of 1,331 hectares. Farm size, however, cannot be directly translated into land ownership. Land holding is more accurately defined as access to land, a system that is based on flexible and highly variable land-leasing contracts similar to other arrangements in more developed countries (Jepson 2003).

Land tenancy also offer insights into the diverse social relations of mechanized production on the Rondônia soybean frontier. Generally farmland in capitalist systems is cultivated under one of three contract forms: (i) ownership (land is cropped by owners), (ii) fixed-rent tenancy (land is cropped by tenant who pays an upfront rent and collects all income), and (iii) sharecropping (land is cropped by tenant who pays a portion or share of the output). Farmer interviews conducted in Vilhena confirm that ownership and sharecropping are the most prevalent tenancy arrangements in the soybean economy. These options are exclusively for capitalist farmers, as peasant farmers, with little credit, limited access to machinery, and no economies of scale, cannot produce enough to pay for soybean production. Fixed-rent tenancy contracts are rarely found in Vilhena.

Soybean farmers sharecrop land through single or multi-year contracts (*arrendamento*), often between four and six years. The payments take two forms, or a combination both forms, over the contract term. Sharecropping contracts are paid for in terms of sacks of soybeans per hectare (3-7

---

[1] IBAMA (Brazilian Institute of Environment and Renewable Natural Resources); EMBRAPA (The Brazilian Agricultural Research Corporation); EMATER-RO (Technical Assistance and Rural Extension Corporation of Rondônia); SEDAM (State Secretariat for Environmental Development)

sacks/ha; 180-420 kg/ha) per year, or less often, in percentages of yield (~10 percent).[2] For reference, the average yield reported in field interviews was calculated at 53 sacks (3,180 kg) of soybeans per hectare. Others sharecroppers agree to renovate old cattle pastures into suitable farmland through the application of fertilizer to increase soil fertility and lime decrease soil acidity. In this arrangement, the sharecropper pays does not pay rent the first two years, and in the third and subsequent years, pays an increasing rent (in sacks of soybean per hectare) until the contract ends. It is also common for one farmer to sharecrop and own land within the same agricultural year.

The importance of the two prominent contract regimes (ownership and sharecropping) is borne out by the number of hectares under cultivation owned and sharecropped by the commercial farmer in Rondônia. For example, the farmer, who owned the least amount of land (68 ha) sharecropped on another 210 hectares. The farm operation with the largest area under cultivation (4,700 hectares) cultivated 2,700 hectares on owned land with the remaining 2,000 hectares as part of a sharecropping arrangement. Of the 13 farmers interviewed, only two cultivated exclusively on land they own, representing only 2,715 hectares; another two cultivated exclusively on rented land, representing only 1,192 hectares.

The authors have identified four categories of mechanized agricultural producer: (i) the sharecropper or fixed tenant farmer, (ii) small holder, (iii) the large holder, and (iv) the corporate or consortium producer. The categories are based upon access to land, labor and capital. The tenant farmers can only access land through sharecropping or fixed-rent contracts. These farmers, who use his own tractors and equipment, do not rely on off-farm labor, and therefore, the scale of operation is limited. They run a substantial risk from year to year because they are heavily indebted. Small farmers will own and cultivate around 300 hectares with the possibility of renting land through fixed-rent or sharecropping contracts. The small farmers rely primarily on family labor and mechanization, with the occasional aid of seasonal labor during peak times. Large farmers cultivate hundreds or thousands of hectares of soybeans, with the intention of expanding his operation. He often owns and rents land for cultivation. The large farmer lives in the nearest urban areas, but not on the farm. A full time foreman, living on the farm with his family, manages day-to-day operations. Although the large farmer makes daily visits to the properties under cultivation, he does not labor on the farm. Rather, he contributes technical expertise to the daily operations and makes decisions on all agronomic aspects of production. The corporate or consortium producers control more than

[2] One sack of soybeans is equivalent to 60 kg.

four thousand hectares. The individuals involved are not farmers but rather investors who hire foremen to manage the day-to-day operations of the farm, in addition to professional agronomists and agricultural engineers who develop farming strategies.

DATA AND METHODS

Satellite remote sensing of land modification, the changes in land characteristics for the same class, demands that the data sources shift from high-spatial resolution, low temporal resolution (e.g., Landsat series and SPOT-HRV) to higher temporal resolution sensors that can monitor at inter-annual scales, such as the Moderate Resolution Imaging Spectroradiometer (MODIS). MODIS data are used to characterize land-cover classes in the temperate and tropical regions; indeed, the 250 m bands were specifically added to the sensor to detect anthropogenic land-cover changes (Townshend and Justice, 1988). MODIS 250 m data have resulted in robust land cover change products, including change detection (Morton et al. 2006; Zhan, 2002), percent forest cover and percent crop mapping (Hansen 2002; Lobell and Asner 2004), general land cover mapping (Wessels 2004), crop mapping (Wardlow 2005; Wardlow and Egbert 2005) and phenology characterization (Wardlow et al. 2006).

Recently, Morton et al. (2006) demonstrated the utility of MODIS 250 m data to identify general land-use types (e.g., cropland, pasture, uncultivated) on recently deforested lands on another Amazonian soybean frontier in the Brazilian state of Mato Grosso. MODIS 250-m data also allow for the detection of subtle changes in land cover using continuous surface attributes at seasonal and inter-annual scales at a moderate spatial resolution (Franklin and Wulder 2002; Galford et al. 2007) using the principles of remote sensing phenology that were first derived using Advanced Very High Resolution Radiometer (AVHRR) image in the 1980s in Africa (Tucker et al. 1984) and have subsequently been used to map land cover (Justice et al. 1985), estimate biomass (Millington and Townsend 1989) and seasonal change in dryland vegetation (Millington et al 1994). Finally, the 250 m spatial resolution of MODIS is sufficient to link farmers' decisions to land-use shifts at a field-scale because individual fields of mechanized farmers are larger than the MODIS pixel size. This analysis has not been possible with AVHRR data because of its 1100 m resolution.

The present study employs a MODIS dataset from a previous study on agricultural intensification in Vilhena, Rondônia from 2001-2005 (Brown et al. 2007). That study used MODIS 250 m Normalized Difference Vegetation Index (NDVI) and Enhanced Vegetation Index (EVI) data to examine, on a near biweekly basis, seasonal plant phenological patterns and inter-annual dynamics of croplands. Previous studies show that seasonal changes in NDVI are related to vegetation phenological stages (i.e., onset of green-up [beginning of growing season], maximum greenness, end of greenness [end of growing season]) (DeFries et al. 1995; Lloyd 1990; Reed et al. 1994). Cropping frequency at the 250 m pixel level is thus possible to determine through basic classification of vegetation index time-series data.

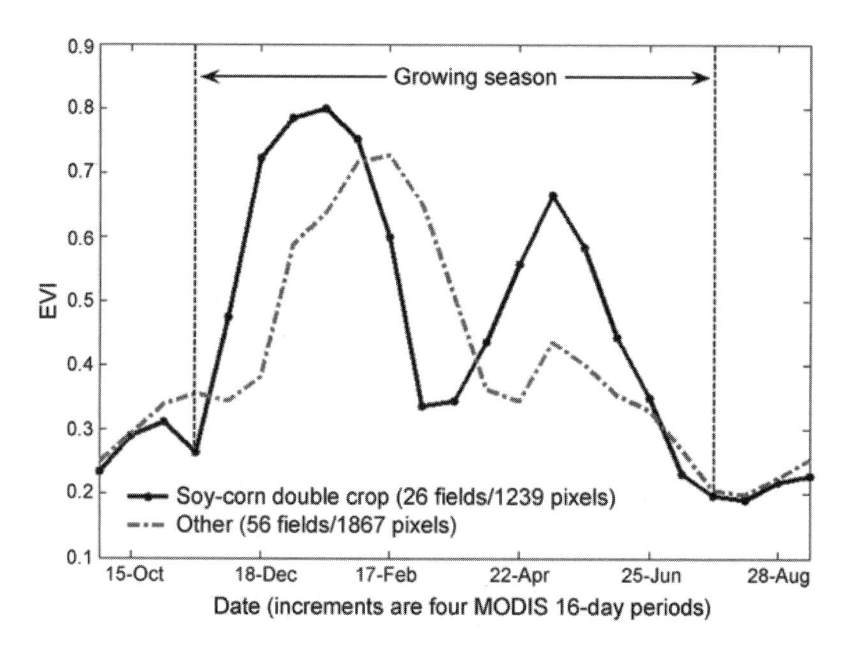

Fig. 5.2. EVI cropping frequency of soy-corn and other fields (Brown et al. 2007: 130)

The data produced by Brown et al. (2007) is a set of annual maps showing the level of agricultural land use intensity across the croplands of Vilhena, Rondônia. MODIS 250 m time-series data provided the required spatial and temporal resolution to detect land modification at the field-scale from one year to the next using a binary hard classification of 'soybean-corn double crop' (SC) and 'not soybean-corn double crop' (not SC) (Brown et al. 2007). Figure 5.2 illustrates changes in EVI are reflective of field-validated cropping practices over one agricultural year. Based on fieldwork and interviews with agronomists at Brazil's national agro-pastoral research agency (EMBRAPA), we confirmed that the most intense land use in Vilhena is a double-cropping regime, defined by one crop of soybeans followed by corn (*safrinha*) during the same agricultural year. All other forms of cropping in Vilhena (e.g., single soybean crop, single rice crop, single soybean crop followed by a cover crop such as millet) represent lower levels of capital investment and inputs than the soybean-corn double crop regime. Therefore, for this study we define land intensification as the transition of a field from management that is 'not soybean-corn double crop' to 'soybean-corn double crop.' Conversely, de-intensification would be from 'soybean-corn double crop' to 'not soybean-corn double crop.'

After creating a mask for the areas of agricultural production under cultivation continuously between 2000 and 2005, we mapped cropping frequency. We identify the crop year as the harvest year, not the planting year. For example, a crop planted in 2000 and harvested in 2001 is identified as 'crop year 2001.' We used the annual maps of cropping frequency ('not soybean-corn double crop'/'soybean-corn double crop') to examine pixel-by-pixel land transitions. This analysis measures intensification and the number of soybean-corn double cropping per pixel over the five year period. For example, the pixel with five soybean-corn double crops (S-C 5) would be considered a land parcel with the highest cropping frequency, and thus highest intensity level. The next highest would be four soybean-corn double crops (S-C 4) over the period. The pixels with the lowest cropping frequency of S-C would be those that register either one or zero over the five-year period (S-C 1 or S-C 0).

RESULTS AND DISCUSSION

Analyses of change in agricultural land-use intensity illustrate three important conclusions about the expansion of soybeans in southern Rondônia. First, there is an upward trend of agricultural intensification in Vilhena (Table 5.2). Overall cropland intensified 37.7 percent between 2001 and 2005, reflecting an upward trend in soybean production for Vilhena. This trend is can be seen in annual changes. Almost 10 percent of land under cultivation experienced intensification between 2001 and 2002, while 22.6 percent intensified between 2004 and 2005. Second, this trend is confirmed when cropping level is examined as a percentage of the overall soybean production. With the exception of crop year 2002, soybeans produced under the soy-corn cropping system (S-C) increases from 30 percent to 39 percent of the total soybean harvest between 2001 and 2005 (Table 5.3).

The land transition analysis (pixel-by-pixel) illustrates that 1,122 hectares were under the highest level of cropping frequency, or five consecutive years of soybean-corn double crop per pixel, while pixels that experienced four years of soy-corn double crop (S-C 4) totaled 1,358.50 hectares (Figure 5.3). The largest area of land (9,790 ha) was under soybean-corn double cropping for only one of the five years of the study. A closer examination of the land-transition results indicate that this follows the trend toward increasing intensification in the region. For example, Table 4 shows that 4,713.50 hectares, or 48 percent of the S-C 1 class, was doubled cropped in 2005. Moreover, the intensification trend is also clear when we examine S-C 2 class. Pixels that were managed under the soy-corn cropping systems in 2004 and 2005 represent 45 percent of the total S-C 2 class area (6,886 ha).

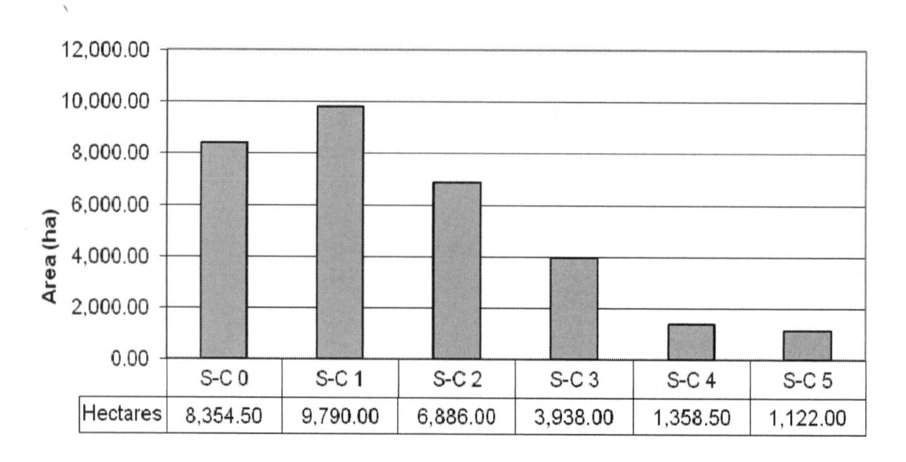

Fig. 5.3. Soy-corn crops per pixel, 2001-2005

Table 5.2. Intensification, Hectares (Percent of Cropland) for Vilhena, Rondônia, 2001-2005

Compared Crop Year	2001/2002	2002/2003	2003/2004	2004/2005	2001/2005
Intensified	3,119 (9.9)	7,772 (24.7)	6,870 (21.8)	7,095 (22.6)	11,847 (37.7)
De-intensified	3,900 (12.4)	2,431 (7.7)	4,356 (13.9)	4,571 (14.5)	2,250 (7.2)
No Change	24,431 (77.7)	21,247 (67.6)	20,224 (64.3)	19,784 (62.9)	17,353 (55.2)

Table 5.3. Area (hectares) of soybean production and soybean-corn (S-C)

Crop Year	2001	2002	2003	2004	2005
S-C Crop[1]	5,038	4,257	9,598	12,111	14,636
Total Soybean[2]	16,700	22,800	29,000	32,000	38,000
Percent S-C of Total Soy Produced in Vilhena	30.17	18.67	33.10	37.85	38.52

Sources: 1. Brown et al (2007); 2. IBGE Municipal Agricultural Production; Area calculated using 5.5ha/MODIS pixel

The transition table also illustrates some agricultural de-intensification although the overall area of de-intensified agriculture is lower than intensification (Table 5.4). For example, 280.50 hectares was doubled cropped for three years (2001; 2002; 2003) and not double cropped for the remaining years. This continued when 412.50 hectares that had been under the soy-corn cropping had been taken out of the system in 2005.

Table 5.4 Land transition for soy-corn double cropping, 2001-2005

2001	2002	2003	2004	2005	Total # S/C	Area (ha)
S-C	S-C	S-C	S-C	S-C	S-C 5	1,122
not S-C	not S-C	S-C	S-C	S-C	S-C 3	2,574
S-C	S-C	not S-C	S-C	S-C	S-C 4	720.5
not S-C	not S-C	not S-C	S-C	S-C	S-C 2	3,124
S-C	S-C	S-C	not S-C	S-C	S-C 4	225.5
not S-C	not S-C	S-C	not S-C	S-C	S-C 2	1,435.5
S-C	S-C	not S-C	not S-C	S-C	S-C 3	720.5
not S-C	not S-C	not S-C	not S-C	S-C	S-C 1	4,713.5
S-C	S-C	S-C	S-C	not S-C	S-C 4	412.5
not S-C	not S-C	S-C	S-C	not S-C	S-C 2	1,133
S-C	S-C	not S-C	S-C	not S-C	S-C 3	363
not S-C	not S-C	not S-C	S-C	not S-C	S-C 1	2,662
S-C	S-C	S-C	not S-C	not S-C	S-C 3	280.5
not S-C	not S-C	S-C	not S-C	not S-C	S-C 1	2,414.5
					TOTAL	31,449

Areas calculated using 5.5 ha per pixel.

We explored the relationship between land tenure and cropping frequency through two analyses: (i) an aggregate analysis and (ii) comparison with land transition categories (Figure 5.4). Field polygons identified as cultivated by owner covered 7,656 ha, and rented lands covered 8,464.50 hectares (16,120.5 total hectares). For the aggregate analysis, we linked agricultural intensity as cropping frequency, the number of harvests in every 250 m pixel of our study area during the 2001-2005 agricultural years, to land tenure. Taking into account length of sharecropping contracts and changing terms (e.g., number of sacks per hectare), we decided to aggregate the number of harvests (or crops) per pixel over a five year period. We also assume, based on field data, that the farmer on any given pixel is the same over the five-year period. We calculated the area of all polygons in the owned and rented category and determined how many crops total over the study period were produced in every 250 m pixel. The total number of harvests produced from the 2001-2005 agricultural years is 17,661 (8,774 crops from owned lands, 8,887 from rented). A chi-square test of the relationship between owned/rented and single/double cropping confirmed no significant relationship.

Second, land transition data were compared with spatial data on land tenure to explore the relationship between intensification levels and land tenure (Table 5.5). Farmers who rented or sharecropped land were early adopters of the soybean-corn double cropping regime. Rented land represents 73.85 percent of the S-C 5 cropland class with tenancy information while only 26.15 percent of the S-C 5 cropland class with tenancy information was cultivated by those farmers who owned land. This temporal trend is also evident in the S-C 4 class. The difference between rented and owned land for the S-C 4 class was 6.2 percent. This trend changes in the S-C 2 and S-C 1 classes. Farmers who cultivate their own land represent a larger percentage of the S-C 2 (52.93%) and S-C 1 (57.38%) cropland class than sharecroppers or fixed-tenant farmers (S-C 2 = 47.07%; S-C 1 = 42.62%).

Table 5.5. Land tenure and cropping frequency, 2001-2005

S-C Class	Rented		Owned		Tenancy Data	Total Cropland
	Area (ha)	% of S-C Class	Area (ha)	% of S-C Class	Area (ha)	Area (ha)
S-C 1	2,304.5	42.62	3,102	57.38	5,406.5	9,790
S-C 2	1,765.5	47.07	1,985.5	52.93	3,751	6,886
S-C 3	1,259.5	55.58	1,006.5	44.42	2,266	3,938
S-C 4	330	53.10	291.5	46.90	621.5	13,58.5
S-C 5	528	73.85	187	26.15	715	1,122
S-C 0	2,277	67.76	1,083.5	32.24	3,360.5	8,354.5
				Total	16,120.5	31,449

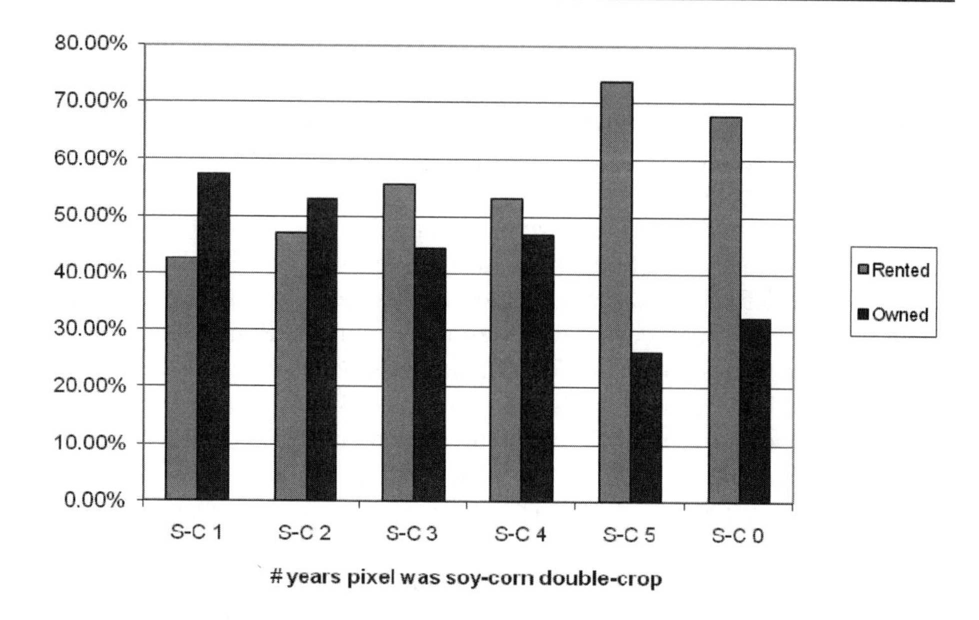

Fig. 5.4 Land tenancy differences among cropping-frequency classes, 2001-2005.

CONCLUSION

Agricultural intensification represents an important land-change dynamic on the Brazilian soybean frontier in Rondônia. The results from this initial analysis provide spatially explicit data supporting the conclusion that the practice of soybean-corn double cropping is increasing. The data show the increasing percentage of regional soybean production under soy-corn double cropping. These findings demonstrate how we may begin to disaggregate land change within agricultural production and identify anthropogenic influence on land modification. Moreover, if paired with data on forest land conversion, these findings will allow for a more complete picture of modern agriculture's imprint on the regional landscape.

Our results also provide one example in which MODIS 250 m data provide important anthropogenic data on inter-annual change, attributable to land management practices. The approach could be expanded into a spatially detailed and reliable measure of agricultural production across the agricultural region of Central Brazil or on other tropical agricultural frontiers, allowing for an assessment of spatial and temporal patterns of variability. This up-scaling would require further in situ validation, as crop green-up and harvest times are highly variable across the production region. However, Vilhena provides an important test case of the approach before any broader scale studies are attempted. Indeed, such data will be of interest to those measuring the biogeochemical footprint of mechanized agriculture (Galford et al. 2007), climate impacts from land-cover change (Feddema et al. 2005), and the broader ecological impacts of agricultural intensification (Matson et al. 1997).

Finally, the analysis provided preliminary results that the spatial resolution of MODIS 250 m datasets is sufficiently detailed to link with socio-economic variables for large-scale agriculture. This is critical in research on the expansion high-input agricultural into tropical landscapes. It allows for land-change scientists to develop new strategies to link current land-cover products with socioeconomic datasets, explain the changes and model land modification and land change. The analysis in this chapter illustrates one way to link variables at the field-level, such as the farmers' land tenure status, with MODIS 250-meter data. While this approach requires further refinement and is restricted to commercial production (with large fields), meaningful results show promise for future studies in the expanding areas of mechanized agriculture and even pasture across tropical South America.

REFERENCES

Alho CJR, and de Souza Martins E. (1995). *De Grão em Grão, o Cerrado Perde Espaço (Cerrado - Impactos do Processo de Ocupaçao)*. Brasília, D. F.: WWF - Fundo Mundial para a Natureza.

Boserup E (1965) The Conditions of Agricultural Growth. Chicago, Aldine City

Browder JO (1994) Surviving in Rondonia: the dynamics of colonist farming strategies in Brazil's northwest frontier. Studies in Comparative International Development 29(3): 45-69

Brown JC (2001) Responding to deforestation: productive conservation, the World Bank and beekeeping in Rondonia, Brazil. Professional Geographer 53(1): 106-191

Brown JC, Jepson W, Kastens JH, Wardlow BD, Lomas JM, Price K (2007) Multi-temporal moderate spatial-resolution remote sensing of modern agricultural production and land modification in the Brazilian Amazon. GIScience and Remote Sensing. 44(2):117-148

Brown JC, Koeppe M, Coles B, Price K (2005) Soybean production and conversion of tropical forest in the Brazilian Amazon: The case of Vilhena, Rondonia. Ambio 6(34): 456-463

DeFries RS, Hansen MC, Townshend JRG (1995) Global discrimination of land cover types from metrics derived from AVHRR Pathfinder data. Remote Sensing of Environment 24:459-479

DeFries RS, Townshend JRG (1994) NDVI-derived land cover classifications at a global scale. International Journal of Remote Sensing 15(17):3567-3586
Fearnside P (2001) Soybean cultivation as a threat to the environment in Brazil. Environmental Conservation 28(1):23-38
Fearnside P, Ferreira GDL (1984) Roads in Rondonia: highway construction and the farce of unprotected reserves in Brazil's Amazonian forests. Environmental Conservation, 11(4):358-360
Feddema JJ, Oleson KW, Bonan GB, Mearns LO, Buja LE, Meehl GA (2005) The Importance of Land-Cover Change in Simulating Future Climates. Science 310(5754): 1674-1678
Franklin SE, and Wulder MA (2002) Remote sensing methods in medium spatial resolution satellite land cover classification of large areas. Progress in Physical Geography 22: 173-2005
Galford GL, Mustard JF, Melillo J, Gendrin A, Cerri CC, Cerri CEP (2007) Wavelet analysis of MODIS time series to detect expansion and intensification of row-crop agriculture in Brazil. Remote Sensing of Environment, in press, available online 17 August 2007.
Galford GL, Mustard JF, Melillo JM (2007). Biogeochemical consequences of land-use transitions along Brazil's agricultural frontier. Retrieved 5 January, 2007, from http://envstudies.brown.edu/reserch/LULCC/research/Brazil.html
Greenpeace International (2006) Eating up the Amazon. Greenpeace International, Amsterdam
Hansen MC, DeFries RS, Townshend JRG, Sohlberg R, DiMiceli C, Carroll M (2002) Towards an operational MODIS continuous field of percent tree cover algorithm: using AVHRR and MODIS data. Remote Sensing of Environment, 83:303-319
Hecht SB, Cockburn A (1990) The Fate of the Forest: Developers, Destroyers and Defenders of the Amazon. HarperCollins, New York
Hecht SB (2005) Soybeans, development and conservation on the Amazon frontier. Development and Change 36(2):375-404
Howden D (2006) Eating the Amazon: The fight to curb corporate destruction. *The Independent*, London
Instituto Brasileiro de Geografia e Estatística. (IBGE) (2005). Produção Agrícola Municipal. Retrieved 1 February, 2007, from http://www.ibge.gov.br
Jepson W (2003) A Political Ecology of Land-use and Land-cover Change in the Central Brazilian Savannas of Eastern Mato Grosso Since 1970. Ph.D. thesis, University of California, Los Angeles
Jepson W (2005) A disappearing biome? Reconsidering land-cover change in the Brazilian savanna. The Geographical Journal 171: 99-111
Jepson W (2006a) Private Agricultural Colonization on the Brazilian Frontier, 1970-1980. Journal of Historical Geography 32(4):839-863
Jepson W (2006b) Producing a modern agricultural frontier: Firms and cooperatives in Eastern Mato Grosso, Brazil. Economic Geography 82(3): 289-316
Justice CO, Townshend JRG, Holben BN and Tucker CJ (1985) Analysis of the phenology of global vegetation using meteorological satellite data. International Journal of Remote Sensing 6:1271-1318

Kaimowitz D, Smith J (2001) Soybean technology and the loss of natural vegetation in Brazil and Bolivia. In: Angelsen A, Kaimowitz D (eds) Agricultural technologies and tropical deforestation. CABI Pub in association with Center for International Forestry Research, New York, 195-211

Keys E, McConnell WJ (2005) Global change and the intensification of agriculture in the tropics. Global Environmental Change 15:20-337

Lambin EF, Geist HJ, Lepers E (2003) Dynamics of land-use and land-cover change in tropical regions. Annual Review of Environment and Resources 28:205-241

Lambin EF, Rounsevell DA, Geist H (2000) Are agricultural land-use models able to predict changes in land-use intensity? Agriculture Ecosystem & Environment 82:21-331

Lisansky JM (1988) Migrants to Amazonia: Spontaneous Colonization in the Brazilian Frontier. Westview Press,Boulder

Liverman D, Moran EF, Rindfuss RR, Stern PC (1998) (eds). *People and Pixels: Linking Remote Sensing and Social Science.* Washington D. C.: National Academy Press

Lloyd D (1990) A phenological classification of terrestrial vegetation cover using shortwave vegetation index imagery. International Journal of Remote Sensing 52:534-544

Lobell DB, Asner GP (2004) Cropland distributions from temporal unmixing of MODIS data. Remote Sensing of Environment 93:412-422

Mahar DJ (1989) Government Policies and Deforestation in Brazil's Amazon Region. The World Bank, Washington, D.C.

Matson PA, Parton WJ, Power AG, Swift MJ (1997) Agricultural Intensification and Ecosystem Properties. Science 277(5325):504-509

Millikan BH (1992) Tropical deforestation, land degradation, and society: lessons from Rondonia, Brazil. Latin American Perspectives 19(1): 45-72

Millington AC, Townshend JRG (1989) Biomass Assessment: Woody Biomass in the SADCC Region. Earthscan Publications: London.

Millington AC, Wellens EJ, Setlle JJ and Saull RJ (1994) Explaining and Monitoring Land Cover Dynamics in Drylands Using Multi-temporal Analysis of NOAA AVHRR Imagery, In Foody G and Curran PJ (eds) Environmental Remote Sensing from Regional to Global Scales. John Wiley & Sons: Christchurch, UK

Morton DC, DeFries RS, Shimabukuro YE, Anderson LO, Arai E, del Bon Espirito-Santo F, Freitas R, and Morisette J (2006) Cropland expansion changes deforestation dynamics in the southern Brazilian Amazon. Proceedings of the National Academy of Sciences of the United States of America, 103(39): 14637-14641

Reed BC, Brown JF, VanderZee D, Loveland TR, Merchant JW, Ohlen DO (1994) Measuring phenological variability from satellite imagery. J of Vegetation Science 5:703-714

Rindfuss RR, Walsh SJ, Turner BL, Fox J, Mishra V (2004) Developing a science of land change: Challenges and methodological issues. Proceedings of the Na-

tional Academy of Sciences of the United States of America, 101(39): 3976-13981.
Schmink M, Wood CH (1992) Contested Frontiers in Amazonia. Columbia University Press, New York
Simon MF, Garagorry FL (2006) The expansion of agriculture in the Brazilian Amazon. Environmental Conservation 32(3):203-212
Smith J, Vicente Cadavid J, Ayarza M, Pimenta de Aguiar J, Rosa R (1999) Land use change in soybean production systems in the Brazilian savanna: the role of policy and market conditions. J of Sustainable Agriculture 15(2/3):95-117
Smith J, Winograd M, Gallopin G, Pachico D (1998) Dynamics of the agricultural frontier in the Amazon and savannas of Brazil: analyzing the impact of policy and technology. Environmental Modeling and Assessment 3(1-2):31-46
Steward C (2007) From colonization to 'environmental soy': a case study of environmental and socio-economic valuation in the Amazon soy frontier. Agriculture and Human Values 24:107-122
Townshend JRG, Justice CO (1988) Selecting the spatial resolution of satellite sensors required for global monitoring of land transformations. International Journal of Remote Sensing 9:187-236
Tucker CJ, Townshend JRG, and Goff TE (1984) African Land-Cover Classification using Satellite Data. Science 227: 369-375
Turner BL II, Brush S (1987). Comparative Farming Systems. Guilford Press, New York
Wardlow BD (2005) The Development of a Crop-related Land-Use/Land-Cover Mapping Protocol in the US Central Great Plains Using Time-Series MODIS 250-meter Vegetation Indices (VI) Data. Unpublished manuscript, University of Kansas, Lawrence, KS.
Wardlow BD, Egbert SL (2005) State-level crop mapping in the US Central Great Plains Agroecosystem using MODIS 250-meter NDVI data. Pecora Symposium series,Sioux Falls, SD.
Wardlow BD, Kastens JH, Egbert SL (2006) Using USDA crop progress data for the evaluation of greenup-onset data calculated from MODIS 250-meter data. Photogrammetric Engineering and Remote Sensing 72:1225-1234
Warnken PF (1999) The Development and Growth of the Soybean Industry in Brazil. Iowa State University Press, Ames
Wessels KJ, DeFries RS, Dempewolf J, Anderson LO, Hansen AJ, Powell SL, Moran EF (2004) Mapping regional land cover with MODIS data for biological conservation: examples from the Greater Yellowstone Ecosystem, USA and Para State, Brazil. Remote Sensing of Environment 92:67-83
Zhan X, Sohlberg RA, Townshend JRG, DiMiceli C, Carroll ML, Eastman JC, Hansen MC, DeFries RS (2002) Detection of land cover changes using MODIS 250 m data. Remote Sensing of Environment 83:336-350

CHAPTER 6 - Coffee Production Intensification and Landscape Change in Colombia, 1970-2002

Andrés Guhl

Centro Interdisciplinario de Estudios de Desarrollo , Universidad de los Andes, Bogotá, Colombia

ABSTRACT

Agriculture is one of the most powerful agents of landscape transformation. The establishment of coffee plantations, and later on, the intensification of production, led to widespread changes in the Colombian Andes. This chapter presents the most important transformations that accompanied the commercial intensification of coffee production in this country starting in 1970. The results indicate that, until 2002, the total area planted in coffee decreased while the area planted with other crops increased This leads to landscape diversification. The adoption of the intensive system also meant the disappearance of most shade trees associated with the traditional coffee production system. At the same time, land tenure has fragmented, the intensive system has favored middle and large land holders, and small holder farmers rely more on wage labor. These results highlight the potential of agricultural intensification in the process of land change in the tropics.

INTRODUCTION

Since the mid-19th Century, coffee has played a very important role in the history and the development of Colombia. As with any other agricultural product, the spread of this crop has transformed not only the rural landscapes of the country, but also the socio-economic conditions of the coffee farmers and the rest of the Colombian people. Because of its importance, the economic, politic, cultural and social impacts of coffee production have been widely studied (Arango 1977, 1982; Bergquist 1978; Beyer 1947; Duque et al. 2000; Escobar and Ferro 1991; Lleras 1980; Nieto 1971; Palacios 1980; Parsons 1968; Ortiz 1989). More recently, the intensification of coffee production has also resulted in widespread landscape transformations. Since landscape change is one of the most important contributors to global environmental change (Mannion 2002; Turner et al. 1995) understanding landscape evolution is a critical step for finding ways to use the resources needed to sustain and improve human well-being while at the same time reducing the human influence on the Earth System (GLP 2005). This chapter describes the landscape evolution of the Colombian coffee lands that results from the adoption of the intensive coffee production system that started in 1970 and discusses some its most important social, economic, and environmental impacts.

AGRICULTURE AND LANDSCAPE CHANGE

Landscapes represent the physical expression of the interaction between human activities and natural drivers of land transformation (Höll and Nilsson 1999). This physical reality is perceived in different ways by the land managers present in any landscape, and these perceptions determine how landscapes and their resources are used in different contexts (Blaikie 1995). Usually, social and economic forces dictate how land is used (Ojima et al. 1994; Rice 1997). Although landscape evolution is a natural process (e.g. succession), it is currently dominated by human activities (Lambin and Geist 2001). Of the entire range of human actions that affect landscape evolution, agriculture and urbanization are the most influential (Brown 2001; Mannion 2002).

Landscape change is a subject that has attracted the attention of the scientific community. For example, the IGBP and IHDP programs have joint research initiative, the Land-Use and Land-Cover Change (LUCC) (Lambin et al. 1999; Turner et al. 1995) and its successor, the Global Land Project (GLP) (GLP 2005), that try to understand the complex dynamics of landscape evolution and its social, environmental and economic implications at a variety of temporal and spatial scales. Much of this research has been restricted to the conversion of one land cover into another, particularly the transformation of forests into agricultural land (Lambin et al. 2000). However, there is a more subtle and gradual process called land cover modification that also leads to significant landscape changes (Lambin et al. 2001; Lambin et al. 2003). For example, secondary succession gradually changes the appearance of a landscape. Similarly, a change in city zoning from residential to other uses will gradually transform the appearance of a particular section of town. Agricultural intensification is also a land cover modification process in which new management practices alter the land cover without necessarily changing its land use (e.g. the land use of an agricultural area undergoing intensification remains agriculture).

It is clear that landscape change is not a simple process; different driving forces act in synergistic ways in different parts of the world to change landscapes (Geist and Lambin 2002). For example, while the clearing of forests for agriculture in Latin America is often the result of large-scale commercial operations, in Africa it is mostly the result of the expansion of the agricultural frontier by small-holder farmers (Lambin et al. 2003). Although in both instances the process involves clearing forest for agriculture, the dynamics and patterns of change are different as a result of different agents of land-cover change.

In the same way, intensification may have several implications for landscapes. Agricultural intensification is the process where the productivity and/or utilization of the land is increased (Netting 1993). In certain areas, where mechanization is possible, agricultural intensification usually leads to more homogeneous and simpler landscapes, and land consolidation (Di Pietro 2001; Hietala-Koivu 2002; Pauwels and Gulinck 2000). In other instances, since intensification leads to higher yields, some land may be abandoned, and intensification may lead to succession in those areas (Southworth and Tucker 2001; Guhl 2004b, 2004a). Finally, intensification at the farm level may lead to diversification of the agricultural land covers present in the landscape (Conelly and Chaiken 2000) and the de-intensification of some land uses (Wiegers et al. 1999). As shown below, the latter is the case in the Colombian coffee lands. These few examples highlight the wide variety of impacts that agricultural intensification and commercialization may have on landscapes. It is expected that these two processes will become increasingly important in the future, as the land available for agriculture reaches some kind of limit, and the increasing demands for food and fiber will require to intensify agricultural production (Conway 2001; Lee et al. 2001; MA 2005). Therefore, understanding the five essential landscape change questions – What has changed? Where has it changed? When did it change? How and why did it change?– caused by intensification is a key priority of landscape change studies and to reducing the human impact on the environment.

COFFEE AND LANDSCAPE CHANGE

The establishment of coffee plantations has very significant impacts on the landscape and the people inhabiting it. For example, the coffee boom in the 19th Century, fueled by the increased demand for the beverage in the rich countries of the northern hemisphere, transformed the human geography and landscapes of tropical America and many mountain slopes were planted with coffee (Price 1994). Changes did not only occur in the appearance and functioning of landscapes, but the social and economic conditions of coffee farmers were also altered. Perhaps the most important change that occurred in the coffee-producing countries is that peasant families that were previously producing mostly for subsistence became more market oriented when coffee was adopted (Parsons 1968). Even in areas not engaged in large-scale agricultural production, coffee became a crop linking the local level with the international market. Agricultural production became the result of land-use decisions based on local conditions and international markets. For many, coffee established a neocolonial relationship between the producers in the former colonies and the consumers in what we know today as the developed world (Topik 1998), and landscapes in the former colonies were related to changes in the patterns of consumption half the world away. At the landscape level, the historic evolution of coffee production was also responsible for major transformations. For example, when production shifted from Haiti to Ceylon in the early 19th Century, it was accompanied by major land conversion in this Asian island. By the time Coffee Leaf Rust (*Hemileia vastatrix* Berk. and Br.) disease wiped out coffee production in Ceylon in the late 1860s, about 71,000 hectares of forest had been transformed into coffee plantations (Dicum and Luttinger 1999). When coffee arrived in Martinique in the early 18th Century, it rapidly became a very popular crop. Within the first three years of its arrival, three million coffee bushes had been planted on the island (Dicum and Luttinger 1999). Assuming a low planting density associated with the traditional coffee production system (2500–3000 bushes per hectare), this represents between 1000–1200 hectares planted just in a very short period of time.

Starting in the late 1960s, the intensification of coffee production also has had significant impacts at the landscape level. In Mexico, the Mexican National Coffee Institute (INMECAFE) promoted the intensification of production, coffee monocultures replaced the more traditional, diverse coffee agriculture (Rice 1996). This was a response to an increasing demand for coffee in the international market. The adoption of this intensive system was mostly the result of technical assistance. The area transformed represented 30 percent of the coffee growing area in Mexico. The intensification of coffee production has the potential to transform more than 5.2 million hectares of coffee fields in Latin America (FAO 2006). The degree to which intensification has been adopted varies drastically from country to country. In 1996 about 70 percent of the fields in Colombia were planted with the intensive production system. In the same year, approximately 90% of the area under coffee in El Salvador corresponded to the traditional shade-coffee production system (Rice 1996).

COFFEE PRODUCTION SYSTEMS

Coffee is planted under several production systems that differ in terms of their vegetational and architectural structure as well as their management. These systems are part of a continuum that ranges from the more traditional systems, where coffee is planted under the canopy of rainforest trees, to the most intensive, which is essentially a coffee monoculture with no shade at all (Moguel and Toledo 1999; Gobbi 2000). In this continuum both the shade trees and coffee bushes are managed. As shade is reduced, there is more space to plant additional coffee bushes. The remaining rainforest canopy may be replaced with a variety of useful tree species, and as one moves towards the more intensive end of the spectrum, shade species become fewer until they disappear. However, it is important to recognize that the idea of planting coffee without shade is not new. The Dutch plantations in Asia in the 18th and 19th Centuries had no shade at all (Dicum and Luttinger 1999), neither did the coffee plantations in Venezuela (Price 1994) and Guatemala (Parsons 1968) in the 19th Century. What makes the intensive system different is the heavy use of external inputs and the adoption of shorter varieties that allow higher planting densities.

Although using shade by itself as a distinguishing characteristic does not capture the full variability of coffee production systems, it has been widely used in the scientific literature as the main factor to distinguish between traditional and intensive coffee production (Niehaus 1992; Ortiz 1989; Palacios

1980; Perfecto et al. 1996; Rice 1997; Rappole, King, and Vega-Rivera 2003; Brown 1996b, 1996a; Dicum and Luttinger 1999; Errázuriz 1986; FNC 2001; Muschler 2001; Baggio et al. 1997; Parsons 1968; Wrigley 1988; Willson 1985). The most important characteristics of these production systems are summarized in (Table 6.1)

The different coffee production systems also have implications for the ecological quality and ecosystem services provided by the coffee agroecosystem. The more intensive systems tend to be less diverse and provide fewer ecosystem services (Ambrecht et al. 2006; Gobbi 2000; Perfecto et al. 2003; Perfecto et al. 1996; Moguel and Toledo 1999). For example, the removal of shade eliminates suitable habitat for a wide variety of species. The presence of shade trees increases the amount of litter on the ground, increases the organic matter content of the soils, and reduces erosion potential. From an economic perspective, the more traditional systems have lower yields than more intensive systems. In the latter, this is the result of higher planting densities and the heavy use of agrochemicals. Therefore, farmers are more dependent on external inputs when they choose the intensive system, and they are (particularly smallholder farmers) more susceptible to market fluctuations and changes in the price of synthetic fertilizers and pesticides (Gobbi 2000). Since the use of external inputs is minimal in the more traditional systems, the traditional coffee production system is regarded as a relatively benign form of agriculture that is well suited for small farms. Although the shade production system is more environmentally friendly than the intensive system, it never replaces the ecosystem services provided by a native forest plot (Rappole et al. 2003).

Table 6.1. Coffee production systems characteristics

Traditional system characteristics	**Intensive system characteristics**
Coffee bushes planted under shade trees	Coffee bushes planted with little or no shade
Low planting densities (<2500 bushes/ha)	High planting densities (up to 10000 bushes/ha)
Traditional (tall) varieties	Short varieties
Little use of inputs	Very high use of inputs
Requires less strict agroecological conditions	Requires more strict agroecological conditions
Long coffee plot life cycle (>10 years)	Short coffee plot life cycle <7 years)

The adoption of the intensive production system has several impacts at the landscape level. Perhaps the most important point is that shade forests have disappeared to a large extent. In most of Latin America, this is important because the shade trees provide ecosystem services and suitable habitat for a wide range of species. Since there are few natural forests left in the coffee growing areas, removing these coffee-associated forests leaves many species without suitable habitat (Perfecto et al. 1996). Another important implication of the intensive production system is that in order to maintain the productivity of the coffee plot, it needs to be replanted or renovated with a higher frequency (FNC 2001). This means the landscape is changing faster than previously, as coffee plots need to be replanted every five years instead of every ten or fifteen as is the case with the traditional system. Additionally, since intensive production system has higher yields, the amount of land needed to attain a certain level of production is smaller than it is under the traditional system. Furthermore, since the agroecological conditions for intensive coffee production are more strict, areas that are not well suited for this kind of production may be abandoned or planted with other crops. Therefore, the adoption of the intensive system may lead to landscapes with smaller areas planted with coffee and larger areas devoted to other land covers, including secondary growth

COFFEE IN COLOMBIA

For much of its history, coffee was the backbone of the country's economy. In 1870, coffee represented 17 percent of the legal exports (Palacios 1980). The importance of this crop increased drastically, and by 1970 coffee was the most important export product, accounting for 63 percent of the total legal export earnings (BANREP 2002a, 2002b). During the 1990s, the importance of this crop decreased in importance as the Colombian economy started to diversify, and oil, gold, and other primary commodities became important sources of foreign currency. By 2000, when coffee prices were very low and production had remained more or less stable for a decade, coffee only represented 8.1 percent of the Colombian legal export earnings. Despite the fact that the importance of coffee in the country's economy has decreased dramatically, in 1997 its production represented the basis of the livelihoods for about four million people, about 35 percent of the agricultural workforce of this Latin American country (FNC 1997).

Historically, coffee in Colombia has been produced in relatively small farms with a wide variety of land uses. (Chalarcá 1998; Escobar and Ferro 1991; Palacios 1980; Parsons 1968). In 1970, the average farm had 3.5 ha devoted to coffee production, representing just 18.5 percent of the area of the farm (FNC 1970). By 1997 the average farm had 1.4 ha of coffee, accounting for 18.2 percent of the total area of the farm (FNC 1997). The combination of small farms and the variety of land covers makes this landscape very patchy (Figure 6.1).

The ideal conditions for coffee production in Colombia are between 1000 and 2000 meters above sea level, with the most productive coffee areas concentrated in the 1200–1800 meter altitude belt (FNC 2001). Because of its location near the Equator, the Colombian rainfall regime exhibits a bimodal distribution that results in two coffee harvesting seasons. The ideal rainfall amount ranges between 1500 mm yr^{-1} in the higher (colder) locations to 2500 mm yr^{-1} in the lower (warmer) elevations (FNC 2001). This means that coffee is planted in the mid-elevation belt of the three mountain ranges that form the Colombian Andes, which occupy roughly the western half of the country.

Fig 6.1. A coffee growing landscape. The view from La Siria on the Road from Manizales to Chinchiná (May 2000)

In other Latin American countries one may encounter the full coffee production system spectrum presented above, but shade by itself may not be a useful criterion for distinguishing between intensive and traditional coffee production systems. However in Colombia using shade is a good way of distinguish between traditional and intensive systems because coffee production has been historically characterized by two production systems: (i) the shade-grown coffee production system, in which the original Andean forest is replaced by a few useful tree species and coffee bushes, and (ii) the intensive coffee production system, in which shade trees have been eliminated or minimized. During much of the history of this crop in Colombia (1850-1970), it was produced under the shade system (Palacios 1980; Parsons 1968).

But starting in the 1970s there was a major transformation in the coffee production system. According to Gabriel Cadena, Director of the National Coffee Research Center (Cenicafé), the thrust for intensive coffee production was the result of phytosanitary conditions rather than economic reasons (Gabriel Cadena, personal communication). By the late 1960s, the coffee leaf rust disease had arrived in Brazil and Central America. Since this disease wiped out coffee production in Ceylon in the 19th Century, the National Coffee Growers Federation was very concerned about the consequences that a similar outbreak could have in Colombia, a country that, at the time, was heavily dependent on coffee. Cenicafé was put in charge of designing a coffee production system that made the crop less susceptible to this disease, and at the same time increased the productivity of coffee production. By the late 1960s, the first experiments were taking place in the department of Caldas (Parsons 1968). These involved reducing or removing the shade, and switching to shorter coffee varieties (*Coffea arabica* L. v. Caturra) in order to be able to fit in more bushes per hectare. For most coffee growers it is a known fact that unshaded plantations have higher yields and a shorter life span (Parsons 1968). Therefore the new system increased productivity not only by increasing the planting density, but also by increasing exposure to sunlight. During the 1970s, Cenicafé designed a coffee production system with associated management practices that include soil conservation measures, fertilizer and pesticide application calendars, and frequencies and methods for plot renovation, among others. This system was strongly promoted by the National Coffee Growers Federation (FNC) agricultural extension service and has transformed the landscapes of the Colombian coffee lands.

THE LANDSCAPE IMPACTS OF THE INTENSIFICATION OF COFFEE PRODUCTION

As with the establishment of coffee plantations, the change in production system had dramatic transformations for the coffee growing landscape and its peoples. Colombia is one of the few countries with relatively detailed spatially explicit data regarding the evolution of coffee production based on three large-scale agricultural surveys carried out by the FNC. In these surveys every coffee farmer was interviewed and every farm was mapped, including both coffee and other land covers for 1970 and 1993-97, and just coffee for 1980 (FNC 1970, 1983, 1997). The general public can access this information at the municipality level where all the farms of a municipality are aggregated. However, there has been very little research on the spatial and temporal dynamics of landscape change, and the impacts of the adoption of this new, intensive production system.

Despite the availability of historic satellite imagery and aerial photographs, there are major limitations for using these sources of information in the Colombian coffee lands, and the coffee surveys are the most reliable data source. The most important limitation is that the mid-elevation ranges of the Colombian Andes are very cloudy. This limits the availability of relatively cloud-free imagery. Furthermore, there are very few images available before 1986. Another key limitation is that coffee farms and plots are small. The median size of a coffee plot in 1997 was only 1.09 ha (FNC 1997), which represents about 12 Landsat TM pixels. This fact, coupled with the rugged terrain that causes patterns of high illumination and shadow, make it extremely difficult to separate areas with coffee from areas under other land covers.

THE REDUCTION OF THE AREA PLANTED WITH COFFEE

The intensification of coffee production has had some profound impacts on the landscape. For the period 1970–1997 the total area planted with coffee decreased by 18.5 percent. At the same time, more than 70 percent of the coffee plots of the country had adopted the intensive system. The number of municipalities engaged in coffee production decreased because the intensive system required more strict agroecological conditions.

Although the area planted under coffee has decreased since the 1970s, the total production has exhibited an increasing trend, reaching a maximum in the early 1990s since which time it has stabilized at around 650,000 tons (Figure 6.2). The decrease in area harvested combined with increased production is a clear sign that intensification has taken place. Yields more than doubled between 1970 and 2000, increasing from 470 kg ha^{-1} to 980 kg ha^{-1} of green coffee (FNC 2002); the national yield continues to rise and, for 2005, it was 1289 kg ha^{-1} (FAO 2006).

At the same time that yields have increased, farms areas have been reduced. This means that the trend towards land consolidation associated with intensification reported in the literature is not present in Colombia. In this case, intensification of production has taken place simultaneously with land fragmentation. One of the possible explanations of this is that mechanization is not possible in the rugged terrain where coffee is produced. Therefore, intensive management in this area is associated more with increased demands for labor and inputs than for machinery. Since the amount of labor that a person can provide is limited and intensive management implies higher labor demands, a smaller farm is easier to take care of.

The decrease in the area planted to coffee has not been homogeneous throughout the country. Although the area devoted to coffee has decreased in 307 of the 514 municipalities that have data for the three surveys, it has also increased in 145 and changed less than 10 percent in 62 (FNC 1970, 1983, 1997) (Table 6.2). This means that coffee production is concentrating in certain areas of the country that, in general, have the best agro-ecological conditions for intensive coffee production.

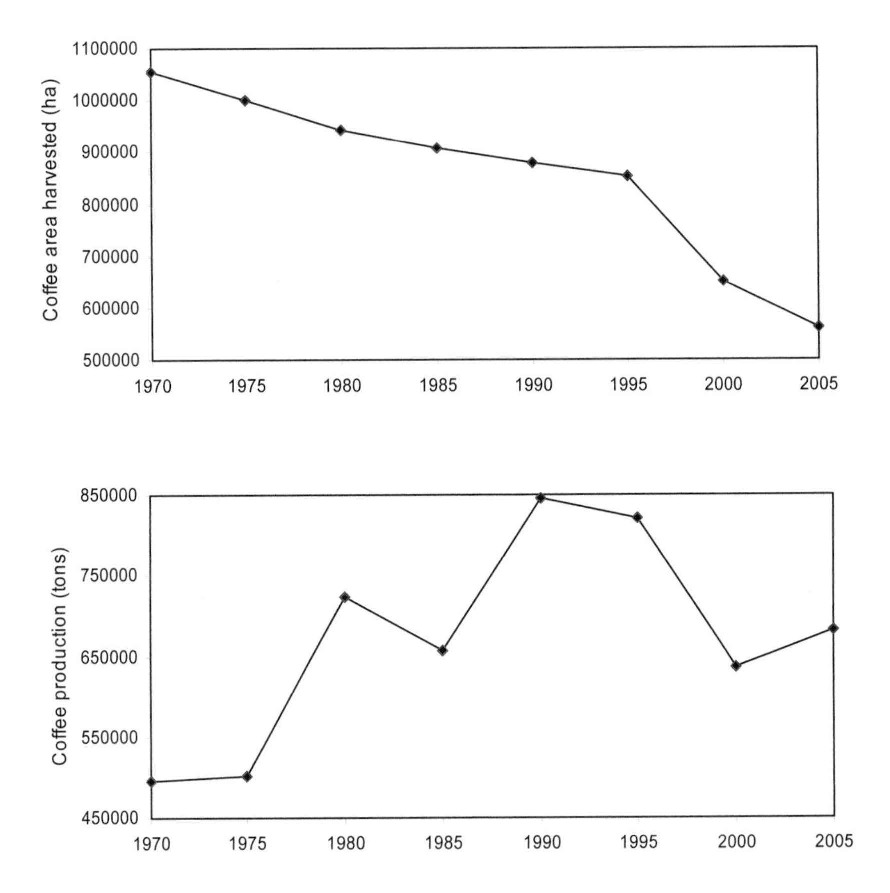

Fig 6.2. Evolution of the coffee area harvested and total production in Colombia between 1970 and 2005 (FAO 2006; FNC 2002)

Table 6.2. Evolution of the coffee growing area, 1970–1997 (FNC 1970, 1983, 1997)

	1970	1980	1993–97
Municipalities surveyed	587	577	559
Number of farms	301,708	n.a.	609,432
Area surveyed (ha)	5,805,732	n.a.	4,773,557
Area planted to Coffee (ha)	1,067,113	1,003,940	869,158
Intensive coffee (% area under coffee)	0.22%	34.14%	70.04%
Coffee area (% of area surveyed)	18.51%	n.a.	18.21%
Forests and secondary growth (ha)	1,026,968	n.a.	1,151,474

DIVERSIFICATION OF THE LANDSCAPE

The decrease in the area planted to coffee between 1970 and 2005 made available land for other crops and land covers. Using information from the large-scale coffee surveys carried out in 1970 and 1993–97, as well as surveys with extension agents carried out in 2002 that present information about land covers in 117 coffee growing municipalities, it is possible to analyze what changes have occurred in the landscape at the municipal level. Six land covers are reported in each survey (Figure 6.3).Four of these are self-explanatory. The other two, 'Other' and 'Temporary' require explanation. Other includes every crop or land cover that does not fit the other five categories. Land covers included in this class are fruit orchards, vegetable production, corn, cassava, and macadamia among others. Temporary refers to the crops that are intercropped with coffee in the early stages of growth. These may include plantains, cassava and, in some instances, corn. Since coffee bushes take about 18 months to start producing (FNC 2001; Willson 1999; Wrigley 1988), Temporary usually refer to unproductive coffee plots.

Since the total area surveyed has decreased as the total area in coffee has diminished, absolute numbers do not give an accurate idea of the composition of the landscape, and how it has changed through time. When the relative contribution of each land cover is calculated, it is evident that the proportion of the landscape planted with coffee has remained nearly constant, while the percentage devoted to pasture has decreased drastically and the share of other crops has increased.

The results also indicate that the forest cover has increased both in absolute and relative terms. Forest cover has increased in eight of the 16 coffee growing provinces (FNC 1976, 1997). Even though it is not clear if forests and secondary succession plots are defined in the same way in the 1970 and 1993-97 FNC coffee surveys, these results are consistent with those reported by Southworth and Tucker (2001) for an area in Honduras. The Honduran study indicates that as coffee production intensification takes place, certain areas of the landscape are abandoned (marginal or isolated plots) and succession takes place. Additionally, the results from Colombia also indicate that the area of forest and secondary growth has increased at the same time as the area in coffee has increased in some provinces. Although the data for Colombia needs to be analyzed with caution since there is no information on the quality of the information and that of the remaining forest cover, the results suggest that coffee production intensification leads to the disappearance of shade forests but it is not necessarily associated with the reduction of natural patches of forest or secondary succession plots.

Unfortunately, there is no information regarding forest and secondary growth at the municipal level. For each municipality it is possible to identify the land covers with the largest increases and decreases (Table 6.3). These results indicate that in 36.7 percent of the coffee growing municipalities pasture has been replaced by other crops, in 19.1 percent of the municipalities coffee replaced pasture, and in 11.9 percent of the municipalities coffee was replaced by other crops. Overall, coffee was the largest land-cover decrease in 18.4 percent and the largest land-cover increase in 25.2 percent of the municipalities. These results emphasize the fact that coffee production is concentrating in certain areas of the country while the total area under production is decreasing.

This table clearly highlights that pasture has been largely replaced by other crops in most municipalities Table 6.3 and Figure 6.3 also suggest that very homogeneous land-covers (pasture) have been replaced by more heterogeneous (other) covers. Therefore, it can be argued that agricultural intensification of coffee production has resulted in a more agro-diverse and heterogeneous landscape, and as land becomes available as a result of intensification other agricultural products appear on the landscape.

Table 6.3. Land cover transitions in the Colombian coffee lands (FNC 1970, 1997)

Percentage of municipalities	Largest land-cover increase1970-97					
Largest land-cover decrease 1970-97	Coffee	Other	Pasture	Sugar cane	Temporary	Total
Coffee	--	11.9%	5.5%	0.8%	0.2%	18.4%
Other	2.5%	--	6.3%	0.6%	1.3%	10.7%
Pasture	19.1%	36.7%	--	1.5%	1.7%	59.1%
Sugar cane	3.1%	2.5%	1.3%	--	0.4%	7.3%
Temporary	0.6%	2.3%	1.5%	0.2%	--	4.6%
otal	25.2%	53.3%	14.7%	3.1%	3.6%	100.0%

The results of the surveys carried out with the agricultural extension agents in 117 municipalities also suggest an increase in landscape agricultural diversity and heterogeneity from 1997, when the last large-scale coffee survey took place, to 2002. Overall, 33 crops increased in importance and 28 crops decreased in importance. In 70 percent of these municipalities there are the same numbers, or more, crops increasing in importance than there are decreasing. This suggests a more agro-diverse landscape is evolving. The adopted crops change from region to region, and a crop which is becoming more important in one area may be decreasing in another. Pasture emerged as a very important land cover as a result of the crisis of in the coffee market between 1997 and 2004. This response has been present during other market swings, and Palacios (1980) argues that during periods of extremely low prices farmers transform coffee plots into pasture

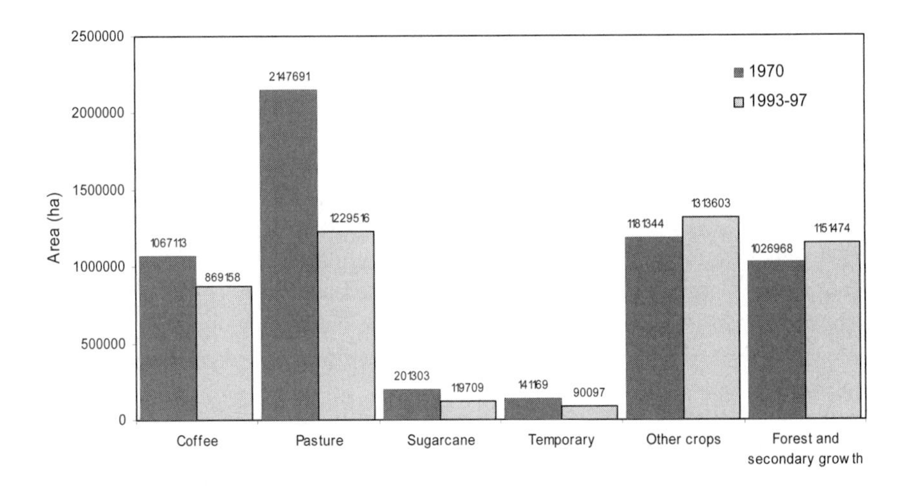

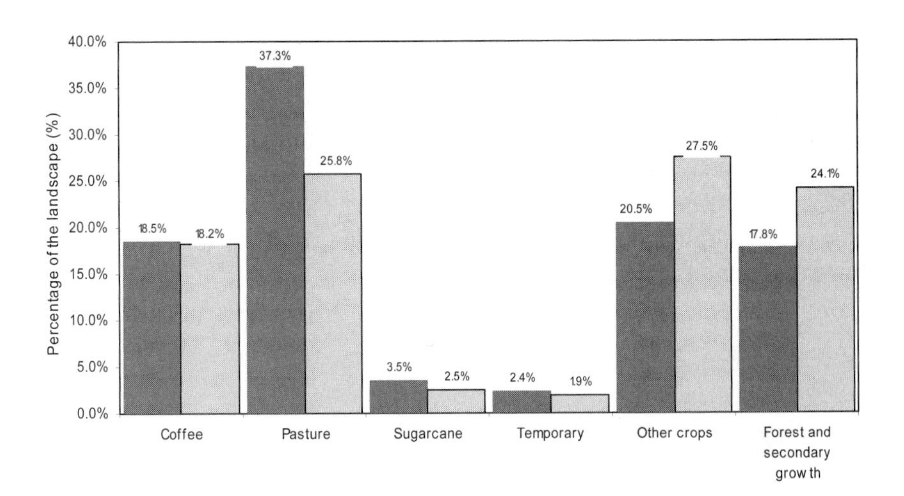

Fig 6.3. Land cover evolution in the Colombian coffee lands (FNC 1970, 1997)

THE NON-LANDSCAPE RELATED IMPACTS OF INTENSIFICATION

The results presented so far clearly indicate that intensification of coffee production has been accompanied by a sharp decrease in the area planted with coffee, and the appearance of new crops in the landscape, making it more agrodiverse and heterogeneous. This transformation has been accompanied with changes in the social, environmental and economic conditions of the coffee lands. This section presents some of the key socio-economic impacts of intensification of coffee production.

One of the most important social and economic impacts has been the differentiation between small, medium and large producers. With the traditional system all coffee farmers work with the same set of rules. But due to the high demand for external inputs of the intensive system, it is better suited to medium and large producers that have better access to capital and credit. Smallholder farmers, which in most instances do not have the capital necessary to buy all the inputs required by the intensive system, tend to maintain the traditional, low productivity production system. Furthermore, they change their livelihood strategies and become wage laborers of the medium and large coffee producers. This transformation has been documented for the municipality of El Líbano, one of the regions of the country where intensification has been widely adopted (Errázuriz 1986), and the results from the 2002 survey indicate that wage labor has become the most important source of income in 105 of 117 municipalities (Guhl 2004b). However, as organic and eco-friendly markets develop and keep growing (Brown 1996a, 1996b; Rappole et al. 2003; SMBC 2003), these traditional coffee plots have the potential to provide once again an on-farm economic alternative to smallholder coffee farmers while at the same time attaining conservation goals. For example, Conservation International has been able to market *Conservation Coffee* with Starbucks and pay higher prices to coffee farmers, who in return have to maintain the shade cover of their coffee plots (Dicker 2002). The International Center for Tropical Agriculture (CIAT) also has a program to help coffee farmers to improve the marketing chain by placing them in direct contact with the consumer, by diversifying their agricultural production and by helping them to become certified in the organic market (CIAT 2006)

Historically, coffee has been responsible for raising the standard of living of the coffee farmers. The National Coffee Growers Federation has invested large sums of money to provide clean water, power, schools, and other kinds of infrastructure in the coffee growing areas. With the recent crises in the coffee market, the FNC was not able to continue with many of its programs. However, since prices have recovered from $0.66/lb 2002 to $1.30/lb at the end of 2006, some of the earnings of coffee are starting to be invested in the well-being of the inhabitants of the coffee growing areas.

Unfortunately, the environmental impacts of coffee intensification are very strong. Historically, the establishment of coffee plantations (traditional system) was responsible for the disappearance of Andean forests and their replacement with useful trees in many areas of the country (Escobar and Ferro 1991; Palacios 1980). In other regions, the adoption of coffee as a crop meant the reforestation of Andean slopes that had been deforested by homesteaders (Parsons 1968). As presented above, the intensification of coffee production resulted in the removal of shade. Additionally, the heavy use of agrochemicals has polluted water courses. The appearance of new crops also means more fertilizers and pesticides are used. According to the surveys with extension agents carried out in 2002, the diversification of agricultural production associated with the intensification of coffee production has also led to more external inputs being utilized, as crops such as vegetables and fruits have higher requirements than coffee. The switch to other non-permanent crops has meant that the soil is exposed for part of the year, thereby increasing erosion rates. In some places the topsoil has been totally lost and production is only maintained with high fertilizer applications (Siavosh Sadeghian, Research Scientist, National Coffee Research Center (Cenicafe), personal communication December 2002).

TOWARDS A CHARACTERIZATION OF LANDSCAPE TRANSFORMATIONS ASSOCIATED WITH INTENSIFICATION

The results presented in this paper clearly indicate the far reaching landscape transformations associated with the intensification of coffee production. These changes not only affect the coffee plot but also influence the other land uses that accompany coffee production. The intensification of coffee production has been accompanied by the diversification of agricultural production and an increase in landscape heterogeneity. Furthermore, most of these products are planted for commercial purposes, and intensification has been accompanied by commercialization of production. Additionally, it would appear that the intensification of coffee production has resulted in the abandonment of some marginal lands that are undergoing succession, highlighting the potential role of agricultural intensification as a tool for environmental restoration. Finally, it is clear that the case study presented for Colombia has social, economic and environmental impacts that go beyond the landscape manifestations of coffee production intensification.

This chapter provides another case study that highlights how intensification of agriculture has a wide variety of landscape impacts. It seems that the outcome of intensification depends on many variables. It is essential to understand how these driving forces interact and lead to different landscape outcomes. Some of these variables include land tenure, possibility of mechanization, and access to markets. In areas with small landholdings and no possibility of mechanization, such as the Colombian coffee lands, intensification has lead to land fragmentation and landscape diversification. The access to markets has also meant that most of the crops present in the Colombian coffee lands are commercial crops. In recent years there have been important attempts to try to characterize how different driving forces and their interactions lead to different deforestation (Geist and Lambin 2002), desertification and dryland degradation (Geist and Lambin 2004) and agricultural change (McConnell and Keys 2005) patterns. Although intensification is one of the components of agricultural change, it is essential to carry out a meta-analysis of land use and land cover change for intensification, since this process will become critical for supplying the increased demands for food and fiber while at the same time minimizing the human footprint on this planet.

REFERENCES

Ambrecht I, Perfecto I, Silverman E (2006) Limitation of nesting resources for ants in Colombian forests and coffee plantations. Ecological Entomology 31 (5):403-410

Arango M (1977) Café e Industria, 1850-1930. 1st edn. Bogotá: Carlos Valencia Editores

——— (1982) El café en Colombia, 1930-1958: producción, circulación y política. 1st edn. Bogotá: Carlos Valencia Editores

Baggio AJ, Caramori PH, Androcioli Filho A, Montoya L (1997) Productivity of southern Brazilian coffee plantations shaded by different stockings of Grevillea robusta. Agroforestry Systems 37(2):111-120

BANREP (2002) Exportaciones No Tradicionales 1970-2001. Banco de la República, Colombia 2002a ,www.banrep.gov.co

———(2002) Exportaciones Tradicionales 1970-2001.Banco de la República, Colombia 2002b , www.banrep.gov.co

Bergquist CW (1978) Coffee and Conflict in Colombia, 1886-1910. Durham: Duke University Press
Beyer RC (1947) The Colombian Coffee Industry: Origins and Major Trends 1740-1940. Ph.D. thesis, University of Minnesota, Minneapolis-Saint Paul
Blaikie P (1995) Changing Environments or Changing Views? A Political Ecology for Developing Countries. Geography 80(348):203-214
Brown LR (2001) Eco-economy: Building an economy for the Earth. 1st edn. New York: Earth Policy Institute, WW Norton & Company
Brown SJ (1996a) Low impact, high interest. Organic coffees: an overview, part II. Tea and Coffee Trade Journal 168(10):32-36
——— (1996b) Organic coffees: an overview, part I. Tea and Coffee Trade Journal 168(9):120-123
Cadena G (2000) Personal communication. Chinchina
Chalarcá J (1998) Vida y hechos del café en Colombia. 1st edn. Santafé de Bogotá: J. Chalarca
CIAT (2006) Cinfo. Information Management for DAPA. Land Use Project CIAT 2006, http://gisweb.ciat.cgiar.org/cinfo/index.php?pageid=4&lang=ENG
Conelly W T, M S Chaiken (2000) Intensive farming, agro-diversity, and food security under conditions of extreme population pressure in western Kenya. Human Ecology 28(1):19-51
Conway G (2001) The doubly Green Revolution: Balancing food, poverty and environmental needs in the 21st Century. In: Lee DR, Barrett CB (eds) Tradeoffs and Synergies: Agricultural Intensification. Economic Development and the Environment, Wallingford: CABI Publishing, pp17-33
Di Pietro F (2001) Assessing ecologically sustainable agricultural land-use in the central Pyrénées at the field and landscape level. Agriculture, Ecosystems and Environment 86:93-103
Dicker A (2002) Personal Communication. Bogotá
Dicum G, Luttinger N (1999) The Coffee Book: Anatomy of an Industry from Crop to the Last Drop. The Bazaar Books Series, New York Press, New York
Duque H, Restrepo M, Velásquez R (2000) Estudio sobre cosecha de café y mano de obra en Palestina, Caldas. Chinchiná: CENICAFÉ
Errázuriz M (1986) Cafeteros y cafetales del Líbano: cambio tecnológico y diferenciación social en una zona cafetera. Bogotá: Universidad Nacional de Colombia
Escobar O L, Ferro G (1991) Cultura del Hombre Cafetero. Pereira: Banco de la República-Area cultural
FAO (2006) FAOSTAT. United Nations Food and Agriculture Organization 2006, http://faostat.fao.org/default.aspx
FNC (1970) Censo cafetero. Anexo. Bogotá: Federación Nacional de Cafeteros de Colombia

——— (1976) Atlas Cafetero de Colombia. Bogotá: División de Investigaciones Económicas, Federación Nacional de Cafeteros de Colombia

——— (1983) Censo cafetero 1980-81. Bogotá: Federación Nacional de Cafeteros de Colombia

——— (1997) Sistema de Información Cafetera. Encuesta Nacional de Cafeteros SICA. Informe Final. Santafé de Bogotá: Federación Nacional de Cafeteros de Colombia

———(2001) Cartilla Cafetera CAFE: Comité de Cafeteros del Valle del Cauca - CENICAFE

———(2002) Infocafé. Asesoría Económica e Internacional -- División de Estudios Especiales y Estudios Básicos

Geist HJ, Lambin EF (2002) Proximate Causes and Underlying Driving Forces of Tropical Deforestation. BioScience 52(2):143-150

——— (2004) Dynamic Causal Patterns of Desertification. BioScience 54(9):817-829

GLP (2005) Global Land Project: Science Plan and Implementation Strategy. Young B (eds) IGBP Report No.53/ IHDP Report No.19,Estocolmo: IGBP Secretariat, Stockholm, pp 64

Gobbi JA (2000) Is biodiversity-friendly coffee financially viable? An analysis of five different coffee production systems in western El Salvador. Ecological Economics 33 (2000):267-281

Guhl A (2004a) Café y cambio de paisaje en la zona cafetera colombiana entre 1970-1997. Cenicafé 55(1):29-44

——— (2004b) Coffee and landscape change in the Colombian coffee lands, 1970-2002. Ph.D. thesis, University of Florida, Gainesville.

Hietala-Koivu R (2002) Landscape and modernizing agriculture: a case study of three areas in Finland in 1954-1998. Agriculture, Ecosystems and Environment 91:273-281

Höll A, Nilsson K (1999) Cultural landscape as subject to national research programmes in Denmark. Landscape and Urban Planning 46:15-27

Lambin EF, Baulies X, Bockstael N, Fisher G, Krug T, Leemans R, Moran EF, Rindfuss RR, Sato Y, Skole D, Turner BL, Vogel C (1999) Land-Use and Land-Cover Change: Implementation Strategy. International Congress of Scientific Unions and International Science Council, Stockholm and Geneva

Lambin EF, Geist HJ (2001) Global land-use and land-cover change: what have we learned so far? Global Change Newsletter, pp 27-30

Lambin EF, Geist HJ, Lepers E (2003) Dynamics of land-use and land-cover change in tropical regions. Annual Review of Enviroment and Resources 28:205-241

Lambin EF, Rounsevell MDA, Geist HJ (2000) Are agricultural land-use models able to predict changes in land-use intensity? Agriculture, Ecosystems and Environment 82:321-331

Lambin EF, Turner, Geist HJ, Agbola SB, Angelsen A, Bruce JW, Coomes OT, Dirzo R, Fisher G, Folke C, George PS, Homewood K, Imbernon J, Leemans R, Xiubin

L, Moran EF, Mortimore M, Ramakrishnan PS, Richards JF, Skånes H, Steffen W, Stone GD, Svedin U, Veldkamp TA, Vogel C, Xu J (2001) The causes of land-use and land-cover change: moving beyond the myths. Global Environmental Change 11(4):261-269

Lee DR, Ferraro PJ, Barrett CB (2001) Introduction: changing perspectives on agricultural intensification, economic, development and the environment. In: Lee DR, Barrett CB (eds) Tradeoffs and Synergies: Agricultural Intensification. Economic Development and the Environment, CABI Publishing Wallingford, pp 17-33

Lleras C (1980) Política Cafetera, 1937/1978. Bogotá: Osprey Impresores

MA (2005) Ecosystems and Human Well-Being Synthesis. Millennium Ecosystem Assessment (MA), Island Press, Washington DC

Mannion AM (2002) Dynamic World. Land-cover and land-use change. Arnold Press, New York

McConnell WJ, Keys E (2005) Meta-Analysis of Agriculture Change. In: Moran EF, Ostrom E (eds) Seeing the Forest and the Trees. Human-Environment Interactions in Forest Ecosystems, MIT Press, Cambridge, MA, pp 325-354

Moguel P, Toledo VM (1999) Biodiversity conservation in traditional coffee systems of Mexico. Conservation Biology 13(1):11-21

Muschler RG (2001) Shade improves coffee quality in a sub-optimal coffee-zone in Costa Rica. Agroforestry Systems 85:131-139

Netting R (1993) Smallholders, householders: farm families and the ecology of intensive, sustainable agriculture. Stanford University Press, Stanford

Niehaus DJ (1992) Slope Instability Hazard Assessment for Natural Disaster Reduction, an approach using Remote Sensing Analysis and Geographical Information Systems: A case Study in the Central Cordillera of Colombia. M.Sc. Thesis, International Institute for Aerospace Survey and Earth Sciences (ITC), Eschende, Netherlands

Nieto LE (1971) El café en la Sociedad colombiana. Bogotá: Ediciones La Soga al Cuello

Ojima DS, Galvin KA, Turner BL (1994) The Global Impact of Land-Use Change. BioScience 44(5):300-304

Ortiz AP (1989) Sombríos y caturrales del Líbano-Tolima. Transformación y crisis ecológica de un paisaje cafetero. Análisis metodológico y cartografía integrada, Vol. 13, Análisis Geográficos, IGAC, Bogotá

Palacios M (1980) Coffee in Colombia 1850-1970. Cambridge University Press, Cambridge

Parsons J (1968) Antioqueño colonization in Western Colombia. 2nd edn. Iberoamericana, University of California Press, Berkeley, CA

Pauwels F, Gulinck H (2000) Changing minor rural road networks in relation to landscape sustainability and farming practices in west Europe. Agriculture, Ecosystems and Environment 77(1-2):95-99

Perfecto I, Mas A, Dietsch T, Vandermeer J (2003) Conservation of biodiversity in coffee agroecosystems: a tri-taxa comparison in southern Mexico. Biodiversity and Conservation 12:1239-1252

Perfecto I, Rice RA, Greenberg R, Van Der Voort ME (1996) Shade Coffee: A Disappearing Refuge for Biodiversity. BioScience 46(8):598-608

Price M (1994) Hands for the Coffee: migrants in western Venezuela's coffee production, 1870-1930. Journal of Historical Geography 20(1):62-80

Rappole JH, King DI, Vega-Rivera JH (2003) Coffee and Conservation. Conservation Biology 17 (1):334-336

Rice RA (1996) Coffee Modernization and Ecological Changes in Northern Latin America. Tea and Coffee Trade Journal 168(9):104-113

——— (1997) The land use patterns and the history of coffee in eastern Chiapas, Mexico. Agriculture and Human Values 14:127-143

Sadeghian S (2002) Personal Communication. Chinchiná

SMBC (2003) Shade Management Criteria for "Bird Friendly" Coffee. Smithsonian Migratory Bird Center 2003, http://nationalzoo.si.edu/ConservationAndScience /MigratoryBirds/Coffee/Certification/criteria.cfm

Southworth J, Tucker C (2001) The influence of accessibility, local institutions, and socio-economic factors on forest cover change in the mountains of western Honduras. Mountain Research and Development 21(3):276-283

Topik SC (1998) Coffee. In: Topik SC, Wells A (eds) The Second Conquest of Latin America. Coffee, Henequen and Oil during the export boom, 1850-1930, University of Texas Press, Austin, TX pp 37-84

Turner BL, Skole D, Sanderson G, Fisher G, Fresco L, Leemans R (1995) Land-Use and Land-Cover Change: Science/Research Plan. International Congress of Scientific Unions and International Science Council, Stockholm and Geneva

Wiegers ES, Hijmans RJ, Hervé D, Fresco LD (1999) Land use intensification and disintensification in the Upper Cañete Valley, Peru. Human Ecology 27(2):319-339

Willson KC (1985) Cultural Methods. In: Clifford MN, Willson KC (eds) Coffee: Botany, Biochemistry and Production. AVI Publishing, Croom Helm Ltd, Westport, pp 157-207

———(1999) Coffee, cocoa and tea. CAB International,New York

Wrigley G (1988) Coffee. In: Wrigley G (ed) Tropical Agriculture Series, Longman Scientific & Technical, Singapore

CHAPTER 7 - Plant Invasions in an Agricultural Frontier: Linking Satellite, Ecological and Household Survey Data

Laura C. Schneider

Department of Geography, Rutgers The State University of New Jersey, Piscataway, NJ 08854-8045, USA

ABSTRACT

Bracken fern has become an important land transformation in the southern Yucatán peninsular region. The fourfold increase from 1985 to 2001 is associated directly with human disturbance, primarily agricultural activities. Once established, bracken fern's persistence is supported by fire, mostly incidental burns from the large amount of swidden fires set every year to clear farm and pasture lands. Its impacts include impediment of forest succession and farmland fallow, reduction in biotic diversity, and high labor costs to combat. This chapter examines the dynamics of bracken fern invasion by linking land-use history, satellite imagery, socio-economic and ecological data through the use of spatially-explicit models. The results of regional and parcel-level models show the importance of richer household survey data and less spatially aggregated socio-ecological data in order to predict the spatial distribution of bracken fern in the region.

INTRODUCTION

An important advance in land-use and land-cover change research is the development of spatially-explicit models aimed at understanding land conversion (Verburg et al. 2006; Veldkamp and Lambin 2001). The improvement in the understanding of the basic geographic and environmental processes associated with land cover change, and the progress in remotely sensed methodologies to characterize landscape have contributed to the advance of spatially-explicit models explaining and predicting processes of land transformation (Lambin and Geist, 2006; Turner, 2002). Land change models provide ways of linking land management decisions to land cover transformations and vice versa, thereby making such models useful at characterizing environmental feedbacks to human decisions, and changes in human behavior to biophysical changes (De Fries et al. 2005; Veldkamp and Lambin 2001).

Most land-change models have focused on understanding and predicting the process of deforestation due to its critical role in the global carbon and hydrological cycles, biodiversity loss, and land degradation more broadly (Foley et al., 2006; Watson et al. 2000). Models of deforestation have primarily focused on documenting the spatial scale and rates of change, and on the social dynamics stimulating forest loss, particularly land management practices. Minimal attention has been given to modeling different kinds of land transitions such as the role of plant invasions in land change (Mooney and Hobbs 2000). So far, models looking at plant invasions affecting the configurations of the landscape are rare (Schneider and Geoghegan 2006) and are mainly found in the ecological literature (Higgins and Richardson 1996; With 2002). Ecological models provide a framework where the spatial character of plant invasion is linked to the physical environment and the biological aspects of the invasive species. Due to the complexities of such relations, the human linkages are less explicit and only mediated through disturbance processes (e.g., fires, land clearance).

Changes in land use, land degradation and human disturbances provide opportunities for invasion, while invading species in turn can force changes in land use or modifications in management (Mooney and Hobbs 2000). Such processes represent a feedback to biophysical change that could be modeled using the spatially-explicit techniques developed by the land change research community. A challenge in modeling plant invasions then is the ability to incorporate social and ecological variables. These linkages can be tested through the spatial characterization of both land cover derived from remote sensing and land use through parcel-level data.

In southern Yucatán, Mexico studies have identified the role of a plant invasive, bracken fern (*Pteridium aquilinum* (L.) *Kuhn*) is an important element of the land-use and land-cover dynamics in the region (Figure 7.1). Bracken has increased fourfold in the area since 1985, impeding regular succession of the vegetation and affecting the spatial configuration of the areas under forest opened for cultivation (Schneider 2004). At first glimpse, bracken fern invasion seems to be the result of land degradation that creates an ecological niche into which bracken can expand. Current spatial configurations of bracken fern invasion in the region, however, suggest a more complex process involving land-use strategies and biophysical constrains (Schneider 2006). In order to understand the variables shaping the current pattern of bracken fern invasion in the region, an integrated approach linking the different socio-economic and ecological factors is required.

Fig. 7.1. Bracken fern mixed with secondary forest in Southern Yucatán

This chapter presents spatially-explicit models of bracken fern invasion in the southern peninsular region of Yucatán that integrate the process of biological invasion with biophysical and socio-economic variables. The models use a spatial dataset, which links together land-use history, satellite imagery, socio-economic and ecological data. The models are developed at both regional and local levels, and some of the challenges when using different scales of analysis are explored. The premise is that bracken fern invasion is an important land change in the region linked to environmental and land management practices, and requires an interdisciplinary framework that integrates the spatial character of the invasion to the understanding of the biophysical constraints and land management practices in the region (Schneider 2004).

SOUTHERN YUCATAN AND BRACKEN FERN INVASION

The study region is located at the frontier of the Mexican-Guatemalan border, an area of 22,000 km^2, which is home of the Calakmul Biosphere Reserve and is therefore considered a hot-spot of tropical deforestation (Figure 7.2) (Lepers et al., 2005). The southern Yucatán region contains the largest tract of continuous deciduous forest in Mexico and it is experiencing an annual rate of deforestation of 0.4 percent, comparable to other forest areas in Central America (Sader et al. 2005). During the Mayan Civilization, the region was intensively used for agriculture but after its collapse the forest returned (Turner 1987). Current forest composition reflects such land-use legacies (Lawrence et al. 2006). At the turn of the 20th Century, new migrants from other regions of Mexico arrived in the region to work mainly in cedar and mahogany logging industries. The increase in population and the ecological importance of the forest endow this region with a continuous conflict between the needs of the farmers and conservation goals.

Currently, the dominant land tenure system in the region is the *ejido*. This sector was created following the Mexican Revolution (1910-1917), a political and social upheaval with its roots in inequitable land distribution. Within *ejido* communities, land is communally regulated by an elected committee; but in this area of southern Mexico, *ejido* members (*ejidatarios*) typically enjoy usufruct access to a single parcel that is permanently allocated to their use.

The main activity of farmers is *milpa* cultivation, a polyculture system combining maize, squash and beans. Following subsistence farming chili peppers are cultivated. They are a cash crop that has proven to be an important source of income for some *ejidos*. Households in the region are also diversifying their off-farm incomes. After cultivation, or during early stages of fallow, grasses are planted on some land parcels in the expectation of cattle ranching. Land clearance either for cultivation or for pasture is one of the main drivers of bracken invasion in the region (Schneider 2006).

Bracken fern invasion affects ecosystem recovery and household economics and it is considered an important part of land-use change in the region (Schneider, 2006; Lawrence et al, 2006). Bracken fern, *Pteridium aquilinum* (L.) Kuhn, has a cosmopolitan plant distribution. It is considered an invasive species because of its tendency to spread out of control, producing a monoculture that discourages the growth of other plant species, thus posing a direct threat to biological diversity (Figure 7.3).

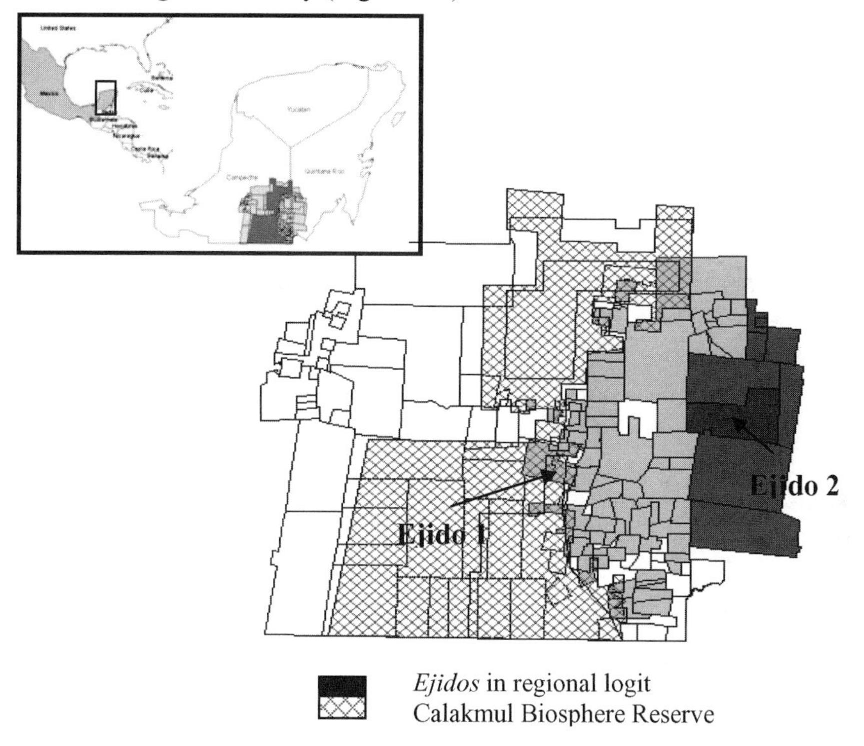

Fig 7.2. Map of study region. Areas in dark grey are the *ejidos* used for regional model. Parcel-level data was used to model bracken fern invasion for *Ejidos* 1 and 2

Bracken fern establishes itself on areas dominated by fires, deforestation, and agricultural activities (Page 1986; Pakeman et al. 1996), causing severe problems to both farmers and conservationists (Pakeman and Marrs 1996, Pakeman et al. 1996). The spread of bracken fern in Southern Yucatan is in part due to land clearance, fire regimes and land management practices (Schneider 2006). The increase of this invasive could potentially lead to further land abandonment and promote greater deforestation in the region (Schneider and Geoghegan- 2006). Thus, specifying where and why bracken fern invades, and why farmers decide to control it or not, are critical components of forest conservation and management in the region.

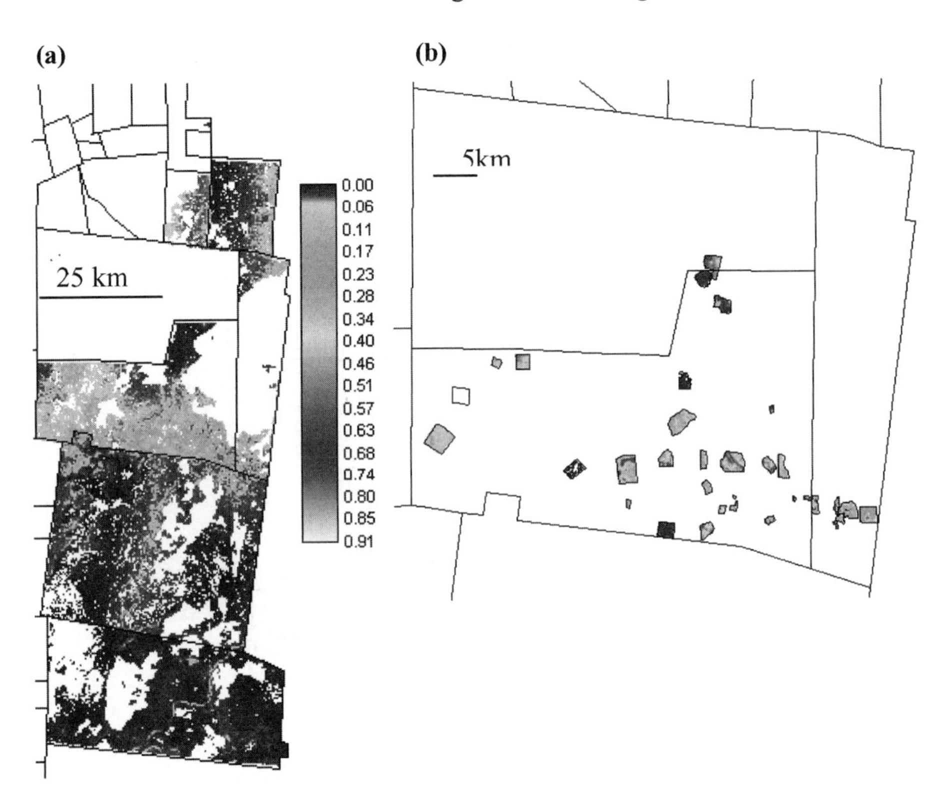

Fig. 7.3 a. Probability of bracken invasion in the study region based on regional model estimations. **b**. Probability of bracken fern invasion using parcel-level socio-economic data

MODELS AND EXPLANATORY VARIABLES

The aims of the land change models presented in this chapter are to understand, through household surveys and census data, some of the dynamics that influence land managers' decisions whether to control bracken invasion and relate those decisions to biophysical constraints through the use remotely sensed data and geographic information systems. Similar approaches have been used to model deforestation in the region (Geoghegan et al. 2004).

Farmers are not directly invading their parcels with fern (as is the case with deliberately cutting forest for cultivation); however, the presence and increase of bracken fern is related to land management practices. As such, links between farmers' land practices and bracken fern invasion are needed to understand the invasion. The model developed in this chapter aims at evaluating such relations and estimating spatially the probability of bracken fern invasion in the region. Two approaches are taken to model bracken fern invasion in the region, one at a regional aggregated level, and the other at a parcel level. The main difference between these two approaches is the quality of the social data used. For the aggregated model, the social data are drawn from the Mexican censuses (from 1990 to 2000), while the parcel-level model employs social data developed directly form the land users for this study (Schneider 2006).

For both the regional and parcel-level approach, the logistic function was used. The logistic function calculates the probability of bracken fern invasion as a function of a set of explanatory variables. Using a spatial data set as the dependent variable allows for spatially-explicit results. The function is a monotonic curvilinear response bounded between zero and one, given by a logistic function of the form:

$$p = (y = 1 \mid X) = \frac{\exp(\sum BX)}{1 + \exp(\sum BX)}$$

where p is the probability of the dependent variable being 1; X is the independent variable (X_1, X_2... Xn); and B is the estimated parameter (B_0, B_1, B_2... Bn).

The logistic function can be transformed into a linear response with the transformation

$$p' = \log_e\left(\frac{p}{1-p}\right) \quad \text{hence} \quad p' = B_0 + B_1X_1 + B_2X_2 + B_3X_3$$

The transformation from the curvilinear response to a linear function is called a logit, or logistic transformation. The logit transformation of dichotomous data ensures that the dependent variable of the regression is continuous, and that the new dependent variable (logit transformation of the probability) is unbounded. Furthermore, it ensures that the predicted probability will be continuous within the range from zero to one (Aldrich et al. 1984).

The dependent variable in each of the models developed in this section is the location and amount of change in bracken fern areas from 1985 to 2001. . The location of those areas was estimated using the land-cover maps created from processing Landsat Thematic Mapper (TM) from 1984-1996 and Enhanced Thematic Mapper (ETM+) for 1999 to 2001. Areas covered by bracken fern are structurally distinct from forest and other disturbed land covers in the southern Yucatan region, which allows them to be detected and differentiated in remote sensing analysis. Using principal components analysis and texture analysis it was possible to separate clearly the spectral signature of bracken fern from other land covers (Roy Chowdhury and Schneider 2004). Maps containing the following eight classes for each time period were created: upland forest, wetland forest, early secondary re-growth, late secondary re growth, agriculture, bracken fern, semi-inundated savannas, and water.

Binary maps of bracken fern-no bracken fern between 1985 and 2001 were created and cross-tabulated to estimate the changes in quantity and the location of bracken fern during that period. The classified maps described above are aggregated into bracken fern and other (including areas of upland forest, agriculture and secondary growth). Areas of lowland forest and inundated savannas are excluded because of the impossibility of these areas supporting bracken fern colonies. The resulting map, with zero indicating absence and one indicating presence, was used as the dependent variable for the models. For both the regional and parcel-level data, the model estimates the probability that a pixel that begins the time period of observation as *not bracken fern* will change into *fern* or will remain as *not fern*. In terms of land tenure, only *ejido* land is used in the model; private land and forest extensions from *ejidos* outside the region and the Calakmul Biosphere Reserve are excluded due to the lack of census data and the differences in tenure.

Explanatory variables

Two sets of explanatory variables are used in the models: biophysical and socioeconomic (Table 7.1). The variables included in the model were chosen looking at how land use might affect the invasion and how environmental variables could constrain or promote the spread of bracken fern in the region.

Table 7.1. Variables used in regional and parcel-level models of invasion

Data	Source	Description
Land cover data	Satellite data: Landsat TM and ETM+ (1987 to 2001)	7 land cover classes, including upland forest, lowland forest, secondary growth, agriculture, bracken fern, inundated savannas and water
Household survey data	Household surveys (n=46)	Demographic data (per household); Land use data, agricultural yields and fallow cycles; and Socio-economic data
Census data	INEGI, 1990 and 2000	Demographic data (per *ejido*); Access to basic services (percentage by *ejido*); and Socio-economic data (by *ejido*)
Ecological data	Digital elevation model, soils and climate from INEGI, 2000	Landscape metrics, slope, elevation and soil type

Data on demography, education, wealth and tenure were selected for the models due to their relationships to bracken fern invasion. Demographic data (population growth, age and gender) provides linkages to amount of land under agricultural production, labor and fern invasion in the land parcels. Population affects control for local demand for agricultural production, which could lead to different land uses (e.g., induced intensification) that constrain or promote the spread of bracken. Education can increase off-farm employment opportunities, thereby leading to an increase in bracken invasion due to land parcel abandonment. Higher levels of education and wealth can have two conflicting effects. On the one hand, wealthier and better educated communities are likely to have more off-farm opportunities. On the one hand, wealthier communities are likely to have more off-farm opportunities, therefore decreasing agricultural activities and increasing the likelihood of invasion. On the other hand, if communities have more social capital, leading to access to subsidies and capital inputs, this could result in intensifying land use and controlling the spread of bracken fern.

The same biophysical explanatory variables are used for both the regional and the parcel-level models. One of the strengths of spatially-explicit modeling is the ability to use geographic information systems and remote sensing to create explanatory variables. Such variables capture the complexity of the landscape and other spatial configurations that affect the process of invasion. The variables included in the regional and parcel-level models for each pixel include: elevation, slope, aspect, fragmentation and soil type. Distance variables and landscape metrics are also included in the model: distance of each pixel to a forested area (zero if the pixel is a forest pixel) and distance of each pixel from paved and secondary roads. In the model, it is assumed that all pixels in each *ejido* have a potential for invasion (excluding seasonal wetlands and areas not considered *ejidos*). The distance of each pixel from the nearest bracken fern pixel was estimated through a map layer that calculated the total number of bracken fern pixels in a 5 x 5 window. It is assumed that the closer to a large area of bracken fern a pixel is, the greater the likelihood of it being invaded.

Socio-economic variables for aggregated regional model

Census data at the *ejido* level is the source for socio-economic data for the aggregated model. The census data used in the model ranges from 1990 to 2000 (INEGI 1990, 2001). The demographic variables used for the regional model were: total population and total number of women and men in each *ejido*. Two variables that include the change in male and female population during this period were also included. Proxy data for wealth are percentage of population with water, sewerage and electricity. Data on education included are percentage of the population that know how to read and write, and the population older than 15 that has attended post-elementary school.

Socio-economic variables for parcel-level model

The sources of the socio-economic data used for the parcel-level model are the surveys developed for this study. A total of 46 (15 for *Ejido 1* and 31 for *Ejido*

2) farmers were interviewed and the parcels were geo-referenced with the use of geographic positioning system (GPS). During each farmer interview, the parcel was walked through and reference points were taken around cultivated plots and boundaries of the parcel. Special GPS points were taken in areas with bracken fern and used to develop training sites for the Landsat TM/ETM+ classification. The results of the visits were a sketch map for each farmer interviewed (Figure 7.3). The spatially-explicit variables developed through GIS and remote sensing were then linked to the parcel through the sketch maps.

Household-level demographic data gathered from the survey included the number of men and women older than 11, and the total number of children under 11. Disaggregating the information by age and gender provides ways to link data on labor and fern invasion in the parcels. Education variables include the education level of the household head as well as the number of household members with more than eight years of education. Because of data limitations, potential wealth and physical capital are included through the percentage of households that have access to basic services and level of middle and high school education.

Variables defining land tenure characteristics are also included in the parcel-level model. The number of years a farmer has been an *ejidatario* and number of years working in the same parcel provides a proxy to the tenure aspects of each parcel. The *ejido* members set the amount of land assigned to each farmer when the *ejido* is founded; the locations of cultivation inside the *ejido*, however, could vary through time. Farmers usually cultivated only the area upon which the *ejidatarios* and the members of the community agreed at the time the *ejido* was founded. *Ejidatarios* vary rarely changes parcels.

In summary, in both models the biophysical and distance variables are the same: slope, aspect, elevation, soils, rainfall, distance of each pixel from primary forest, number of bracken fern pixels in a 5 x 5 window, size of the *ejido*, pixels belonging to a forest extension and distance of each pixel from paved and secondary roads. The socio-economic data for the regional model aggregated at the *ejido* level (Figure 7.1) drawn from the census includes: total population in 1990, total female and male population in 1990, number of residents with access to piped water and electricity and the number of people older than 15 with more than elementary school education. For the parcel-level, the socio-economic data comes from a household survey and it includes: the numbers of males and females older than 11, the number of children, the number of years the head of the household was at school, the number of years as *ejida-*

tario, the number of years working in the parcel visited and vehicle ownership by a household member.

REGIONAL AGGREGATED MODEL

The regional aggregated model was estimated for the eastern part of the southern Yucatán peninsular region using a logit function explained above. This sub-region extends from the Calakmul Biosphere Reserve eastward. The eastern section has larger areas invaded and it was the focus of much of the fieldwork underpinning this research. Table 7.2 shows the main land cover transitions occurred in the sub-region between 1985 and 2001 and it shows how the increase in bracken fern areas is larger than the increase in areas used for agricultural production.

The estimated coefficients and statistical significance of each of the variables used in the model are shown in Table 7.3. Figure 7.2(a) shows the probability of bracken fern invasion based on the model estimations. Due to collinearity among the census data variables, a few variables mentioned in the previous section were dropped from the analysis. The signs of the coefficients are interpreted given the statistical significance as follows: a positive coefficient means an increase in the probability of bracken fern invasion and a negative sign a decrease in the probability.

Table 7.2. Land cover estimates for the southern Yucatán peninsular region (nine *ejidos*: total area: 2,138 km^2)

Year	Bracken fern	Agriculture	Secondary vegetation	Lowland forest	Upland forest
1985	15.8	93.2	203.7	434.6	1391.0
1994	40.1	90.4	187.2	421.9	1420.0
2001	75.0	67.9	190.8	416.1	1389.0

All socio-economic variables resulted in a negative correlation with bracken fern invasion. In terms of demographics, an increase in the population will decrease the probability of bracken fern invasion. Change in total population from 1985 to 1990 and the total number of men in the population has a negative relation with the invasion. The result supports the hypothesis that increasing land pressures provide an incentive to combat the spread of the fern and a larger labor force provides the strength to do it through weeding.

Positive correlations exist between some of the biophysical variables and the invasion. The greater the slope angle, the higher the probability of invasion; this results reflects the inability of the species to grow in seasonally inundated terrain which, in this area, has low gradients. Parcels with steeper slopes are not usually suitable for agriculture therefore farmers would not spend time removing areas of bracken. The invasion by bracken of steep slope areas is most likely driven by fires originated in areas nearby. The relationship between rainfall and soils and bracken is positive. The former reflects the preference by farmers for rendzinas for agricultural purposes. Once vegetation is cleared the first step to facilitate the invasion has been taken.

The spatially-explicit variables show some interesting results. First, a longer distance from forest and a more fragmented landscape are positively correlated with bracken fern invasion, indicating how cleared land is preferred by bracken for establishment and spread. Distance variables and the fragmentation index suggest that an area surrounded by and in close proximity to forest withstands bracken fern invasion better than other areas, either owing to the nearby repository of plant species that could compete with bracken fern invasion or to the protection from fire afforded by the forest. Areas close to roads have higher probabilities of invasion, in part because incidental fires commonly burn along roadsides. Forest extension land has a lower probability of invasion, an expected result given that land is protected from being cleared in the extension zone.

The spatial predictions showing the probability of invasion are shown in Figure 7.3 (a). The best way to evaluate how good the predictions are is to compare them with the actual event. There is not a clear way to compare predicted values with actual values; however, in the actual event land is either invaded or not, and the predicted value represents a probability value. The prediction is a probability of change, while the actual is discrete--either the pixel changes or not. To move the probability to a prediction, a critical value is chosen and values at either side of it determine the result.

A two-fold approach is taken here to assign the critical value. The actual amount of invasion is calculated using the remotely sensed data; then the spatial distribution of higher probability pixels are sequentially identified until the actual amount has been selected. Using IDRISI Kilimanjaro GIS software, the probability image from the logit model was ranked. The highest ranking pixels were selected and assigned as predicted fern. The critical value for selecting the areas predicted as invaded in the model was 0.28.[1]

Table 7.3. Estimated coefficients and standard errors from the regional logit model

Dependent Variable: Bracken invasion or no invasion	**Estimated coefficient**	**Standard error**	**t - statistic**
Socio-economic			
Forest Extension (0,1)	-4.935	0.08649	-57.06
Total population (per *ejido*)	-0.0107	0.00128	-8.33
Change in total population from 1990 to 2000 (per *ejido*)	-0.0024	0.00039	-6.16
Change in male population from 1990 to 2000 (per *ejido*)	-0.0030	0.00039	-7.71
Literate population (per *ejido*)	0.0603	0.00422	14.29
Higher education (per *ejido*)	-0.0289	0.00133	-21.75
Ejido size (# pixels, hundreds)	-1.7e-05	3.89e-07	-45.77
Distance			
Distance to road (m)	-0.00024	2.04e-06	-118.40
Distance to nearest primary forest (m)	0.00462	0.00003	142.35
Biophysical			
Fragmentation index (0,1)	2.60080	0.04364	59.59
Soils	0.71361	0.02055	34.71
Elevation (m.a.s.l.)	-0.00388	0.00020	-19.27
Rainfall (mm/yr)	0.00292	0.00011	26.43
Aspect (0 to 360°)	-0.00030	0.00004	-7.90
Slope (degrees)	0.12851	0.00140	91.43
Number agricultural pixels (5x5)	-0.08285	0.00221	-37.37
Constant	-4.19483	0.16181	-25.92
Pseudo R^2	0.2213		
Number of observations	2,415,730		

The models do not attempt to predict the actual amount of the bracken fern invasion in the region. Taking the amount as given, the model predicts the spatial distribution of the area invaded over the landscape. A map of the areas selected by the model to be invaded (predicted bracken invasion) is then com-

[1] Critical values used for models are usually 0.5, meaning that probabilities of 0.5 or higher will be usually chosen to predict an event.

pared to the map of areas that have actually been invaded by bracken fern during the period of 1985/87 to 2001.

Figure 7.3 shows the probability map of invasion for the eastern part of the region, and Figure 7.4 shows the results of spatial cross-tabulation. Of the pixels that were predicted to be fern, 34 percent were actually fern and only 2 percent of the pixels predicted as fern were actually non-fern (Table 7.6). The fact that most of the area in the model is non-fern enhances the accuracy to predict correctly the pixels kept as non-fern. Even though less than half of the pixels were predicted correctly as fern, it is important to notice that the mistakes are usually clustered in space and close to the correctly predicted, invaded areas.

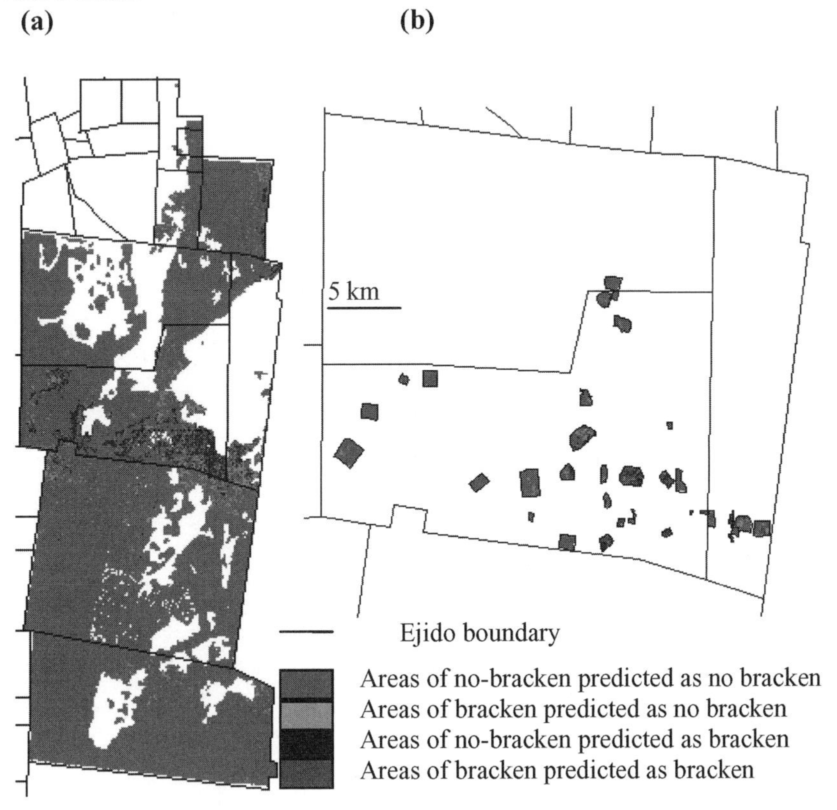

Fig. 7.4. a. Results of spatial cross-tabulation between predicted values from regional model and actual areas cover by bracken fern in 2001. **b.** Cross tabulation for *Ejido* 1 using the predicted-values from parcel-level data

The poor results obtained in predicting the spatial location of invasion could be attributed to the level of aggregation, mainly for the socio-economic variables. The variables used as proxies of demographics and wealth are less successful in capturing a very diverse range of land-use management strategies and histories that exist in the region. Models of deforestation in the region using similar data set also show low levels of prediction. When modeling deforestation, a regional model of deforestation correctly predicts 33 percent of the deforested pixels; for the pixels that remain forested, the regional model correctly predicts 96 percent of those pixels (Geoghegan et al. 2004).

Table 7.6. Results of Ordered Least Square regression model at parcel level.

<table>
<tr><th></th><th colspan="2">EJIDO 1</th><th colspan="2">EJIDO 2</th></tr>
<tr><th>Dependent variable:
percent change in bracken</th><th>Estimated coefficient</th><th>Prob>|t|</th><th>Estimated coefficient</th><th>Prob>|t|</th></tr>
<tr><td>Socio-economic (varies by household (hh))</td><td></td><td></td><td></td><td></td></tr>
<tr><td>Number of men > 11 (per hh)</td><td>0.158</td><td>0.058</td><td>-0.039</td><td>0.023</td></tr>
<tr><td>Number of children <12 (per hh)</td><td>-0.060</td><td>0.149</td><td></td><td></td></tr>
<tr><td>Education of household head in years (per hh)</td><td>-0.107</td><td>0.043</td><td>-0.026</td><td>0.006</td></tr>
<tr><td>Number of years since arrived (years)</td><td>0.009</td><td>0.305*</td><td>-0.006</td><td>0.022</td></tr>
<tr><td>Working days per weeding (per hh)</td><td>-0.014</td><td>0.060</td><td></td><td></td></tr>
<tr><td>Distance and landscape (varies by hh)</td><td></td><td></td><td></td><td></td></tr>
<tr><td>Distance from household to plot (per pixel) (km^2)</td><td>-0.016</td><td>0.062</td><td>-0.038</td><td>0.003</td></tr>
<tr><td>Distance from parcel to nearest road (per pixel) (km^2)</td><td>-3.5E-05</td><td>0.07</td><td></td><td></td></tr>
<tr><td>Distance from parcel to primary forest</td><td>0.001</td><td>0.039</td><td>-0.002</td><td>0.014</td></tr>
<tr><td>Biophysical (varies by hh)</td><td></td><td></td><td></td><td></td></tr>
<tr><td>Upland soil (0,1)</td><td>0.685</td><td>0.05</td><td></td><td></td></tr>
<tr><td>Slope (degrees)</td><td>0.092</td><td>0.053</td><td></td><td></td></tr>
<tr><td>Elevation (m.a.s.l.)</td><td></td><td></td><td>0.001</td><td></td></tr>
<tr><td>Constant</td><td>-0.380</td><td>0.351</td><td></td><td>0.399*</td></tr>
<tr><td>Number of observations</td><td>31</td><td></td><td>15</td><td></td></tr>
<tr><td>F (10, 20)</td><td>3.24</td><td></td><td>(6,8)11.50</td><td></td></tr>
<tr><td>Prob > F</td><td>0.0121</td><td></td><td>0.0015</td><td></td></tr>
<tr><td>R^2</td><td>0.6185</td><td></td><td>0.8961</td><td></td></tr>
</table>

PARCEL-LEVEL MODEL

A logit model and ordered least square (OLS) multivariate regression were estimated at the parcel level. The results of the logit model are used in this chapter to compare modeling approaches at different scales of analysis: parcel-level and regional model (or aggregated model). The OLS approach provides ways that compare the relationships between explanatory variables and bracken fern spread among *ejidos* with contrasting land management practices. Such comparison is difficult with a regional approach because the *ejido* level data is aggregated.

The dependent variable in the logit model is the same as in the aggregated model: pixels in the parcels that change into bracken or not from 1985 to 2001, the unit of observation being the pixel. The differences in the parcel-level model are in the independent variables, which come from the household surveys. For the OLS regression model, the dependent variable is the percentage change in amount of area under bracken fern from 1985 to 2001 in each parcel surveyed, the unit of observation in this case being the number of parcels (*Ejido* 1 = 15, *Ejido* 2 = 31).

The results of OLS are not spatially explicit, but allow testing different statements regarding land management practices and bracken invasion. Examples of the statements are: bracken fern invasions are more severe where larger plots or many contiguous plots have been cleared and burned frequently. Bracken fern is most successful in areas where farmers have sufficient land such that they can afford to lose agricultural plots to the invasive. Farmers who are land constrained are more likely to manage the invasion in order to minimize the spread of the fern.

For spatially-explicit results a logit regression was estimated using the pixels in each parcel as the dependent variables. Using the same set of variables as the regional model, the logit model estimates the probability of a pixel being invaded and the results are presented in a spatially-explicit manner (Figure 2b). For the OLS and logit models, calculations were made to evaluate if the explanatory variables were statistically significant. The larger number of observations of the logit model resulted in all explanatory variables being statistically significant in contrast with the regression model where just few were (Tables 7.4 and 7.5).

Table 7.4. Results of logit model at the parcel level for *Ejido* 2

Dependent variable: bracken or not bracken	Estimated coefficient	Standard error	t -statistic
Socio-economic (varies by household (hh), n=31)			
Number of men > 11 (per hh)	0.878	0.384	22.87
Number of children < 12 (per hh)	-0.134	0.013	-10.21
Education of household head in years (per hh)	-0.261	0.023	-11.33
Number of years since arrived (years)	0.027	0.002	9.29
Number of years worked on the same parcel (per hh)	-0.029	0.002	-10.18
Distance and landscape (varies by pixel)			
Distance from household to plot (per pixel) (km^2)	-0.055	0.036	-15.27
Distance from parcel to nearest road (per pixel) (km^2)	-0.0001	9.39e-06	-17.63
Distance from parcel to primary forest	0.003	0.0001	17.03
Biophysical (varies by pixel)			
Soil	1.652	0.118	13.93
Slope (degrees)	0.076	0.007	10.88
Constant	-4.151	0.166	-24.93
Number of observations	5164		
Pseudo R^2	0.1719		

In terms of spatial predictions at the parcel level, the same method used for selecting pixels in the aggregated model was used to choose the pixels more likely to be invaded from the logit model. The maps of probability and validation for *Ejido* 2 are shown in Figure 7.3, and Table 7.5 shows the results of the cross-tabulation between the predicted and actual areas of bracken fern for *Ejido* 2: 54 percent of the pixels predicted as invaded were actually invaded, and 11 percent of the pixels that were predicted to be bracken were actually not invaded. The critical point for pixel selection was 0.34, higher than in the aggregated model, illustrating the impact of improved scalar congruency in the data.

The fact that at the parcel-level socio economic information is better captured could indicate the improvement of the results over the regional aggregated approach. In terms of prediction, this model seems to represent better the patterns of invasion at the parcel level. A larger sample would improve the results of the models making it possible to include a larger number of explanatory variables that could be evaluated.

Table 7.5. Cross-tabulation of actual with and without bracken fern (2001) and areas predicted by the regional and parcel-level logit models

Regional Model	Actual area without bracken (km^2)	Actual area with bracken (km^2)	Total
Predicted area without bracken	1502.6	45.90	1548.5
Predicted area with bracken	45.9	23.86	69.76
Total	1548.5	69.76	1618.3
Parcel-level Model			
Predicted area without bracken	17.14	2.15	19.29
Predicted area with bracken	2.15	2.49	4.64
Total	19.29	4.64	23.93

There are several demographic and economic differences between the two *ejidos* used to estimate the OLS regression. The first is the fact that bracken fern density is larger in *Ejido* 2 than in *Ejido* 1. A larger number of children and men older than 11 reside in *Ejido* 1 than *Ejido* 2, and a slightly greater number of women per household live in *Ejido* 2 than in *Ejido* 1. The average number of years of education of the heads of the households is higher in *Ejido* 2 than in *Ejido* 1. The average years in tenure are longer in *Ejido*2—almost double the years of the farmers in *Ejido*1. Distance measurements indicate that cultivated plots in *Ejido* 2 are closer to farmers' houses than in *Ejido*1 (In *Ejido* 1 farmers are at least 10 times farther away from their plots than farmers in *Ejido* 2). Also parcels in *Ejido* 1 are closer to forested areas than in *Ejido* 2. Distance to roads in both *ejidos* is similar, however. Yields for the year 2000-2001 in *Ejido* 1 are lower than in *Ejido* 2, where weeding is more frequent. Elevations are higher and slopes greater in *Ejido* 1, and annual average rainfall is slightly higher in *Ejido* 2. Overall, *Ejido* 2 has a larger population but a smaller number of children than *Ejido* 1; as in much older *Ejido* 2, established as a forestry effort, land holdings are larger and land has been used longer compared to *Ejido* 1.

The total number of independent variables used for the OLS regression I less than that used for the logit model because of the sample size (in order to have enough degrees of freedom for the statistical estimations). The variables were chosen for the model as follows: two independent regressions were estimated, one only with socio-economic variables and the other with biophysical-distances variables. Only statistically significant variables were retained. For *Ejido 1*, a total of six independent variables were included in the model, and for *Ejido 2* a total of 10 (Table 7.6). From the household survey these were: the number of men older than 11, the number of children (younger than 11), the number of family members with high school education, the number of years they have lived in the area, the number of years working in the parcel surveyed, and the distances of the parcels from roads and their houses. The biophysical variables used were slope angle, soil type and distance from primary forest. The distance measures were the same those used in the regional model, with the addition the distance between the parcel and their houses in the villages.[2] The results of the OLS regression (Table 7,6) show the relationship between the increase in bracken and socio-economic characteristics. The calculations were made to evaluate if the explanatory variables are statistically significant and how they differ by *ejido*.

The regression model estimations presented here allow the selection of statistically significant variables. First, interpretation of the relationship between the explanatory variables and bracken fern invasion are discussed for *Ejido* 2. All the demographic variables included in the model with the exception of time of arrival in the *ejido* were statistically significant. An increase in percentage of bracken change with the increase in the number of men older than 11 per household and (the not statistically significant) the longer a head of household has lived in the region. A decrease in the amount of bracken fern in a parcel occurs with an increase in the number of children, the number of years of education of the head of household and greater the number of years they have worked the same parcel. The probability of invasion is less if parcels are farther away from roads; but is increases if parcels are farther way from primary forest, the parcels are located on rendzinas and the slopes are steeper.

---

[2] In both *ejidos* the area assigned to each *ejidatario* was set at the time the *ejido* was founded. For *Ejido* 2 each farmer has access to 100 ha; in *Ejido* 1 only 40 ha. The size of the plots worked at the time of the survey visit varied among *ejidatarios*, however.

Some of the results are consistent with expectations. Families later on in their lifecycle and residency tend to have a large set of parcels and can afford to lose some to invasion. However, those earlier in the lifecycle tend to work longer to control the spread of bracken. Parcels close to roads are usually cleared and accidentally burned more, which leads the fern colonies to establish and spread. Cleared areas distant from forest are also more vulnerable to fires and the colonization of secondary vegetation more difficult as the repositories of forest species are further away. The *a priori* expectation on the education variables is that greater education increases the off-farm employment opportunities, thereby leading to an increase in the presence of bracken due to the abandonment of land.

A counter intuitive result, perhaps, is the positive relationship between the number of men in a household and fern invasion in *Ejido 2*. Characteristics of land management in the region show that extra labor for weeding and cutting results in better control of the spread of bracken; consequently, having more men in the household helping in those activities should result in better control of bracken fern. The result found may reflect the lifecycle issue noted above and the fact that the older households in *Ejido 2* tend to have surplus land and may be moving from farming into off-farm activities (e.g., services).

The results of the regression model for *Ejido 1* are a bit different from *Ejido* 2 (Table 7.6). None of the biophysical variables were statistically significant.[3] The most interesting difference is a negative relationship that exists between number of men older than in a household and fern invasion. *Ejidatarios* in this *ejido* have less land and are more dependent on farming activities for their income than *ejidatarios* in *Ejido* 2. The contrasting results indicate that labor availability operates in connection with the orientation of the household economy.

COMPARING LOCAL VERSUS REGIONAL MODELS OF BRACKEN FERN INVASION

The difference in probabilities between the regional-census logit model and the parcel-level logit could be considered a measure of the improved fit of the

[3] Elevation was statistically significant in the regression when only biophysical variables were included. For this reason, elevation was included in the final regression model.

household model, given the higher cut-off probability. For the aggregated model, the critical point was 28 percent and for parcel-level model, 35 percent. Another way to evaluate the relative fits of the models is to compare the total number of correct and incorrect predictions. The regional logit model correctly predicts 34 percent of the invaded pixels, while the parcel level correctly predicts 54 percent of the pixels affected by bracken invasion. For the pixels that remain as forest, the regional logit model correctly predicts 97 percent of those pixels, while the parcel level correctly predicts 89 percent of those pixels.

The improvement of the parcel-level model for predicting invasion reflects the value of adding the richer household level survey data that better captures the individual causes of invasion than is possible at the aggregate level with the census data in the regional model. An additional possible reason for the better predictions of pixels that remain forest for the regional logit model over the parcel model is that for the latter the agricultural plot boundaries for the individual household are known from the sketch maps associated with the household survey. For the regional model, however, only the boundaries of each *ejido* are known, not the boundaries of the potential agricultural land within each *ejido*. As much *ejido* land is designated as communal forest and is therefore ostensibly off limits for cultivation; the model coincidentally achieves a relatively high accuracy in correctly predicting such land not to be invaded.

As with other attempts to model land-use and land-cover change, there remains much data to be collected and much modeling to be done to capture the variation and dynamics of the invasion in the region. As previously discussed, less aggregated socio-economic data, e.g. information on the location of within-*ejido* settlement, would likely improve the accuracy of the regional model. Further modeling work includes adding more temporal observations to the models and, for the parcel-level model, an increase in the number of parcels included could help in such improvements.

CONCLUSION

The human-environment dynamics that give rise to invasive species are an important element of landscape dynamics, with consequences ranging from impacts on biodiversity to net primary productivity and human use values. For the most part, research on plant invasions has examined these plant species in terms of the biophysical dimensions alone with only modest links to the human dimensions. Importantly, the spatial dimensions of invasion are not necessarily a component of their analysis. This chapter provides an example of how socio-economic data at different levels of aggregation, ecological and remotely sensed data could elucidate the dynamics of plant invasions through the use of spatial explicit models.

The models combine different types of spatial data and different levels of spatial aggregation. Some of the dynamics of bracken fern invasion and the variations of spatial patterns are understood through spatially-explicit models that move beyond modeling the magnitude of changes within a region to modeling the locations of that change.

Biophysical and spatial data hint that certain environmental aspects (e.g., aspect, fire frequencies and concentration of limiting nutrients create differences in susceptibility to bracken establishment, and repeated burning of fern-dominated areas favors its retention. Socio-economic and spatial data, however, indicate that the willingness of farmers to combat bracken fern invasion is related to the land, labor, and capital conditions of the farm household. Characterization of a spatially-explicit physical environment and land-use practices result in understanding current process of disturbance and resource distribution in the region which contribute to the spread of bracken and determine its current pattern.

One of the applications of such models is the ability to forecast trajectories of invasion, which are difficult to assess due to the complex and unpredictable nature of the relations among environmental responses and the changing political and social conditions of the systems affected by plant invasions. Plant invasion could be better understood if models could estimate the relative roles of land-use strategies and ecological responses.

REFERENCES

Aldrich JH, Nelson FD (1984) Linear, Probability, Logit, and Probit Models., Sage University Publication, Newbury Park

De Fries R, Asner GP, Houghton R (2005) Trade-offs in Land-Use decisions: Towards a Framework for assessing multiple ecosystem responses to land-use change. In: De Fries R, Asner GP, Houghton R (eds) Ecosystems and Land-use change. American Geophysical Union, Washington DC, pp 1-9

Foley JA, De Fries R, Asner GP, Barford C, Bonan G, Carpenter SR, Chapin FS, Coe MT, Daily GC, Gibbs HK, Helkowski JH, Holloway T, Howard EA, Kucharik CJ,

Monfreda C, Patz JA, Prentice IC, Ramankutty N, Snyder PK (2005) Global Consequences of Land-use. Science 309:570-574

Geoghegan J, Schneider LC, Vance C (2004) Temporal Dynamics and Spatial Scales: Modeling Deforestation in the Southern Yucatan Peninsular Region. Geojournal 61(4):353-363

Lepers E, Lambin EF, Janetos AC, DeFries R, Achard F, Ramankutty N, Scholes RJ (2005) A Synthesis of Information on Rapid Land-cover Change for the Period 1981-2000. Bioscience 55:115-124

Higgins SI, Richardson DM, Cowling RM, Trinder-Smith TH (1999) Predicting the landscape-scale distribution of alien plants and their threat to plant diversity. Conservation Biology 13(2):303-313

Lambin EF, Geist HJ, Rindfuss RR (2006) Local Processes with Global impacts. In: Lambin EF, Geist HJ (eds) Land-Use and Land-Cover Change: Local processes and Global impacts. The IGBP Series, Springer, pp 1-8

Lawrence D, Vester H, Perez-Salicrup D, Eastman R, Turner II BL, Geoghegan J (2006) Integrated Analysis of Ecosystem Interactions with land-use change: The Southern Yucatán Peninsular Region. In: De Fries R, Asner GP, Houghton R (eds) Ecosystems and Land-use change. American Geophysical Union, Washington DC, pp 277-291

Mooney HA, Hobbs RJ (2000) Invasive species in a changing world. Island Press, Washington DC

Page CN (1986) The strategies of bracken as permanent ecological opportunist. In: Smith RT, Taylor JA (eds) Bracken: Ecology, land-use and control technology. Parthenon Publishing Group Press, pp 173-180

Pakeman RJ, Marrs RH, Howard DC, Barr CJ, Muller RM (1996) The bracken problem in Great Britain: its present extent and future changes. Applied Geography 16: 65-86

Pakeman RJ, Marrs RH (1996) Modelling the effects of climate change on the growth of bracken (Pteridium aquilinum) in Britain. Journal of Applied Ecology 33:561-575

Roy Chowdhury R, Schneider LC (2004) Land-Cover/Use in the southern Yucatán peninsular region, Mexico: Classification and Change Analysis. In: Turner II BL, Geoghegan J, Foster D (eds) Integrated Land-Change Science and Tropical Deforestation in the Southern Yucatán: Final Frontiers, Oxford Geographical and Environmental Studies, Clarendon Press of Oxford University Press, Oxford, pp 105-141

Sader SS, Roy Chowdhury R, Schneider LC, Turner II BL (2004) Forest Change and human driving forces in Central America. In: Gutman G, Janetos AC, Justice CO, Moran EF, Mustard JF, Rindfuss RR, Skole D, Turner II BL, Cochrane MA (eds) Land-Change Science: Monitoring, and Understanding Trajectories of Change on Earth's surface. Kluwer Academic Publishers,Dordrecht, Netherlands, pp 57-76

Schneider LC, Geoghegan J (2006) Land Abandonment in an Agricultural Frontier after a Plant Invasion: The Case of Bracken Fern in Southern Yucatán, Mexico. Agricultural and Resource Economics Review 35(1):167-177

Schneider LC (2006) The effect of land management practices on bracken fern invasion in the region of Calakmul, Mexico. Journal of Latin American Geography 5(2):91-107

Schneider LC (2004) Bracken Fern (Pteridium aquilinum (L.) Kuhn) Invasion in Southern Yucatán Peninsular Region: A Case for Land-Change Science. Geographical Review 94(2):229-241

Turner II BL (1983) Once Beneath the Forest: Prehistoric Terracing in the Ri'o Bec Region of the Maya Lowlands. Westview Press, Boulder, CO

Turner II BL (2002) Toward Integrated Land-Change Science: Advances in 1.5 Decades of Sustained International Research on Land-Use and Land-Cover Change. In: Steffan W, Jäger J, Carson D, Bradshaw C (eds) Challenges of a Changing Earth: Proceedings of the Global Change Open Science Conference, Springer-Verlag, Heidelberg, GR, pp 21-26

Verburg P, Kok K, Pontius RG, Veldkamp A (2006) Modeling Land-Use and Land-Cover Change. In: Lambin EF, Geist HJ (eds) Land-Use and Land-Cover Change: Local processes and Global impacts. The IGBP Series, Springer, pp 117-135

Veldkamp A, Lambin EF (2001) Predicting land-use change. Agriculture, Ecosystems and Environment 85:1-6

Watson RT, Noble IR, Bolin B, Ravindranath NH, Verardo DJ, Dokken DJ (2001) Land-use, Land-Use Change and Forestry. Special Report of the IPCC (Intergovernmental Panel of Climate Change), Cambridge University Press, Cambridge

With K (2002) The Landscape Ecology of Invasive Spread. Conservation Biology 16(5):1192-1203

CHAPTER 8 - Shifting Ground: Land Competition and Agricultural Change in Northern Côte d'Ivoire

Thomas J. Bassett and Moussa Koné

Department of Geography, University of Illinois, Urbana-Champaign, IL 61801-3671, USA

ABSTRACT

This paper examines changing land use patterns and heightened competition over land in northern Côte d'Ivoire between 1984 and 2004. Land use is diversifying from an earlier emphasis on cotton and food crops to one that now includes tree crops, notably cashews and mangoes. Rather than abandoning cotton, farmers are extensifying production by spreading costly agricultural inputs over a larger than recommended area. This shift from agricultural intensification to more extensive cotton growing and tree planting is changing the countryside. Land competition is manifest in the expansion of cotton fields and orchards at the expense of rangelands, in land disputes among customary land managers, and in the monetization of customary land lending practices and grazing rights. The paper situates these new land use and land competition dynamics in the context of (1) economic and political crises that have made cotton growing increasingly unattractive, and (2) impending changes in how land is held in Côte d'Ivoire following the passage of a new land law in 1998. The flurry of tree planting and competing land claims reveals how individuals and groups are positioning themselves in relation to the uncertainties ushered in by the new agrarian (dis)order. The case study of the Katiali region illustrates the social and economic dynamics of land use and land cover change at the local scale and how these are related to wider national and international processes linked to neoliberal economic reforms.

INTRODUCTION

Cotton stands out as the most important cash crop in the savanna zone of West Africa. Côte d'Ivoire has until recently been a leader among *franc zone* cotton producers who today account for 15 percent of world cotton exports.[1] This chapter focuses on the shifting ground of cotton cultivation of Côte d'Ivoire in the context of economic and political crises. The shift is from a rural economy based on the intensification of cotton to a more extensive pattern of cotton growing and the widespread planting of tree crops. We use the word *crisis* in the sense of a turning point. In the case of Côte d'Ivoire, the institutional structures and conditions of cotton growing and marketing changed so significantly between 1990 and 2005 that the old order that formed the basis of the cotton economy no longer holds. The new order is fraught with such instability and uncertainty that farmers are modifying their farming systems.

The main sources of agrarian disorder and new land use practices are the neoliberal economic policies of the World Bank and the 2002 rebellion that divided the country into a rebel-controlled north and a government-held south. The rebellion grew from a failed coup d'état in September 2002. Since the cotton growing areas lie entirely within the rebel-controlled zone, the ensuing political instability has greatly disrupted the cotton sector (see below). The cotton economy was already in disarray following the break-up of the parastatal cotton company (CIDT) in the 1990s and early 2000s in the context of World Bank structural adjustment programs. These programs aimed to reduce government budget deficits by eliminating price supports and subsidies to farmers and by reducing the role of the state in the so-called private sector. The World Bank's neoliberal economic reforms were premised on the idea that privatization would result in higher producer prices as a result of greater competition among cotton companies. The belief that privatization would be a boon to cotton growing contained optimistic assumptions about prices for both cotton and agricultural inputs. But the period 1990-2005 has been one in which producer prices have greatly fluctuated while the prices farmers pay for criti-

---

[1] The franc zone countries are former French colonies in west and central Africa whose currency, the CFA franc (FCFA), is tied to the Euro at a fixed rate of 655 FCFA to 1€. These countries include Benin, Burkina Faso, Côte d'Ivoire, Mali, Niger, Senegal, Togo, Cameroon, Central African Republic, and Chad.

cal agricultural inputs have risen. Cotton growers have responded to a general worsening terms of trade on a number of fronts, including boycotting markets, defaulting on agricultural input loans, organizing unions, and even constructing their own cotton gins (Bassett 2001, 146-73). This chapter focuses on the strategy of farmers to sustain income levels through agricultural extensification and diversification.

The dynamics of agricultural change, especially the planting of tree crops, are also linked to the anxieties and maneuvers of individuals and groups on the eve of rural land privatization in Côte d'Ivoire. Promoted by the World Bank in the context of its green neoliberal reforms in the global South (Goldman 2005), the 1998 land law requires the registration and titling of all rural lands. Land privatization is expected to induce greater investments in land and result in higher levels of productivity and environmental conservation (Bassett, et al. 2003). The process of land registration begins in rural communities where individuals must first establish their land rights. Individuals and groups are currently positioning themselves to make such claims by planting trees since tree planting presupposes and signifies long-term rights in land. Thus, by planting cashew and mango trees, people are staking claims to land with the hope of securing title to it in the future. The stakes are high not only for securing farmland but also a cash income. For example, the managers of lineage lands now require immigrant farmers to pay cash and in-kind fees in order to farm. Individual field owners also insist that immigrant livestock producers pay a fee before their animals graze the stubble of harvested fields. This ability to use land as a source of cash will become more widespread as individuals secure land title.

In summary, multiple and multi-scale processes are reshaping the rural economy of northern Côte d'Ivoire. To keep afloat in the teetering cotton sector, farmers are extensifying production by spreading their increasingly costly inputs over a larger area. Farmers with sufficient capital and land rights are investing in tree crops and cattle, and exacting land rents from immigrants. These tactics are producing tensions within and between communities over land access, control, and use. This paper examines these local frictions within the larger political economy (Tsing 2005), and considers their implications for rural land use and livelihoods in the current agrarian (dis)order.

The case study of Katiali is used throughout to explore these trends in land use and land cover change. This village of some 2300 people is at the center of a 570 km^2 area that is recognized by its neighbors as constituting the *terroir* or village lands of Katiali. Senufo farmers were the first settlers in the area. They became the region's ritual land managers by being the first people to clear land, a process that involves ritual sacrifices to the bush spirits residing there. Subsequent migrants must seek permission from one of the more than dozen lineage heads, village chief, or land priest (*tarfolo*) who manage this social and natural heritage for future generations.

The Katiali area lies within the Sudanian climatic and vegetation zone characterized by a bimodal rainfall regime that ranges between 1200–1300 millimeters per year (César 1991, 55-58; Dibi 2004). The savanna vegetation is dominated by bush savanna, cropland, grass savanna and different types of savanna woodlands (gallery forests along rivers, tree savannas and open woodlands). Population grew at a mean annual rate of 3.1 percent between the 1984 and 1998 censuses. The immigration of farmers and herders from the regional centers of Mbengué (pop. 6878) to the north and Korhogo (pop. 142,039) to the south, and from the neighboring countries of Mali and Burkina Faso, are major components of population growth.

THE COTTON ECONOMY IN COTE D' IVOIRE

Cotton is the major cash crop in the savanna region of Côte d'Ivoire (Figure 8.1). It is grown by farmers who cultivate food crops for household consumption and cotton for export. Cotton typically accounts for 40–45 percent of total crop area in the cotton growing areas and is cultivated as a monocrop. Food crops such as maize, rice, yams, millet, sorghum and peanuts cover the remaining farm area. The size of cotton fields varies between one and five hectares. Households relying on manual labor cultivate around one hectare of cotton while households using ox plows plant two to five hectares in cotton.

From a sociological standpoint, Côte d'Ivoire cotton growers are petty commodity producers (Bernstein and Woodhouse 2001, 302). They have access to land through lineage-based land rights systems and rely mainly upon household labor in farming. Cotton-growing households depend upon market relations to meet their basic needs (food, school fees, health care, farm equipment) and have internalized commodity relations in ways that influence their farming systems. In cases of worsening terms of trade, peasant farmers may seek to maintain previous income levels by intensifying production or by reducing consumption levels, or both simultaneously. This *simple reproduction squeeze*, which is experienced as increasing costs of production and decreasing returns to labor (Bernstein 1978), is resisted by cotton growers in northern Côte d'Ivoire. Through a process of agricultural extensification and diversification, cotton growers seek to reduce the costs of production and increase returns to labor. However, these income-earning strategies depend on an abundant land base and sufficient land rights. The relative scarcity of these resources and rights is expressed by competition over land and changes in land cover.

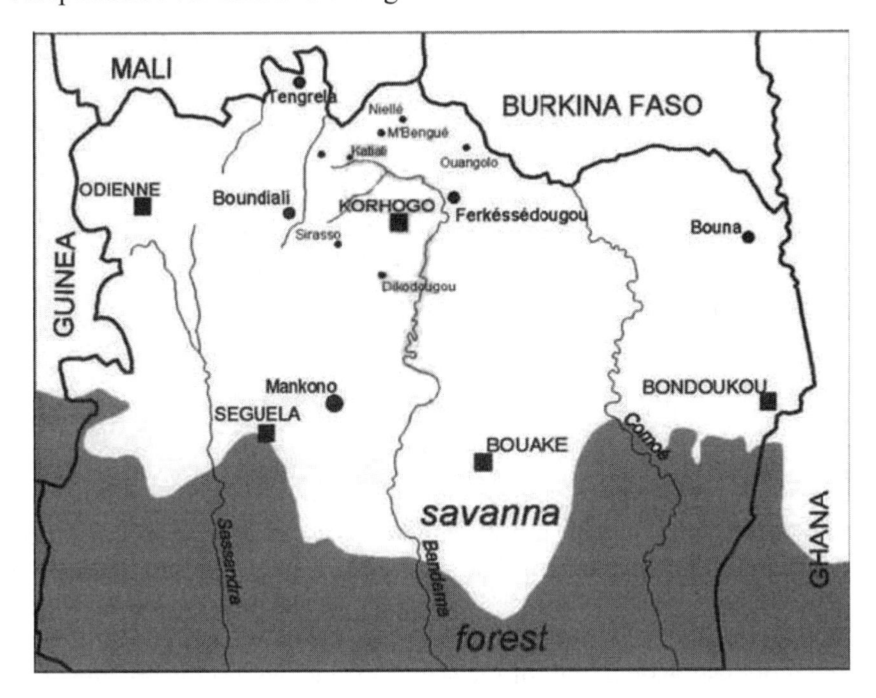

Fig. 8.1. Katiali region in northern Côte d'Ivoire

The intensification of cotton, 1965-1984

Between 1965 and 1984, seed cotton production rose 35-fold in Côte d'Ivoire, rising from 6,000 to 212,000 tons (Figure 8.2). Much of this growth was due to the expansion in cotton area which increased at a rate of 17 percent per year. This expansion was tied to a three-fold increase in the number of cotton growers and a near doubling in size of cotton farms associated with the spread of ox plows. But improved yields were also an important factor in this agricultural success story. Cotton yields increased at an annual rate of 4 percent between 1965 and 1984. Overall, productivity increases accounted for 15 percent of the growth in seed cotton production over this period.

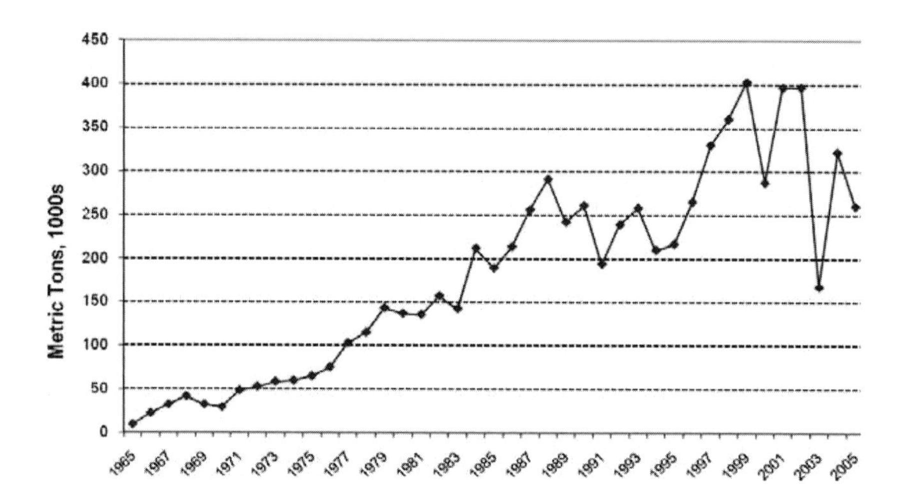

Fig. 8.2. Cotton production in Côte d'Ivoire, 1965-2005

It is impossible to separate the cotton revolution in West Africa from the institutional structures and social relations that tie producers to cotton companies in arrangements similar to contract farming. These institutional structures are most clearly defined by their vertical integration in which cotton companies play pivotal roles in input supply, extension, and marketing. The monopsonistic nature of the sector means that producers must sell their cotton to the company or gin that controls their area. There are some variations on this institutional structure but generally speaking, cotton growers' expenses and income are directly tied to their structural relationships with cotton companies (Figure 8.3).

Prior to 1998, the Ivorian cotton economy was dominated by the parastatal CIDT owned 70 percent by the Côte d'Ivoire government, and 30 percent by CFDT, the former colonial cotton company that changed its name to Dagris in 2001.[2] CIDT was the exclusive buyer and seller of all Ivorian cotton. It provided agricultural inputs to producers on credit, purchased all cotton grown in the country at a fixed price, transported, ginned, and sold cotton fiber and cotton seed, and maintained rural roads and extension agents in the cotton growing areas.

[2] CFDT refers to *La Compagnie française pour le développement des fibres textiles*. Dagris is *Développement des Agro-industries du Sud*.

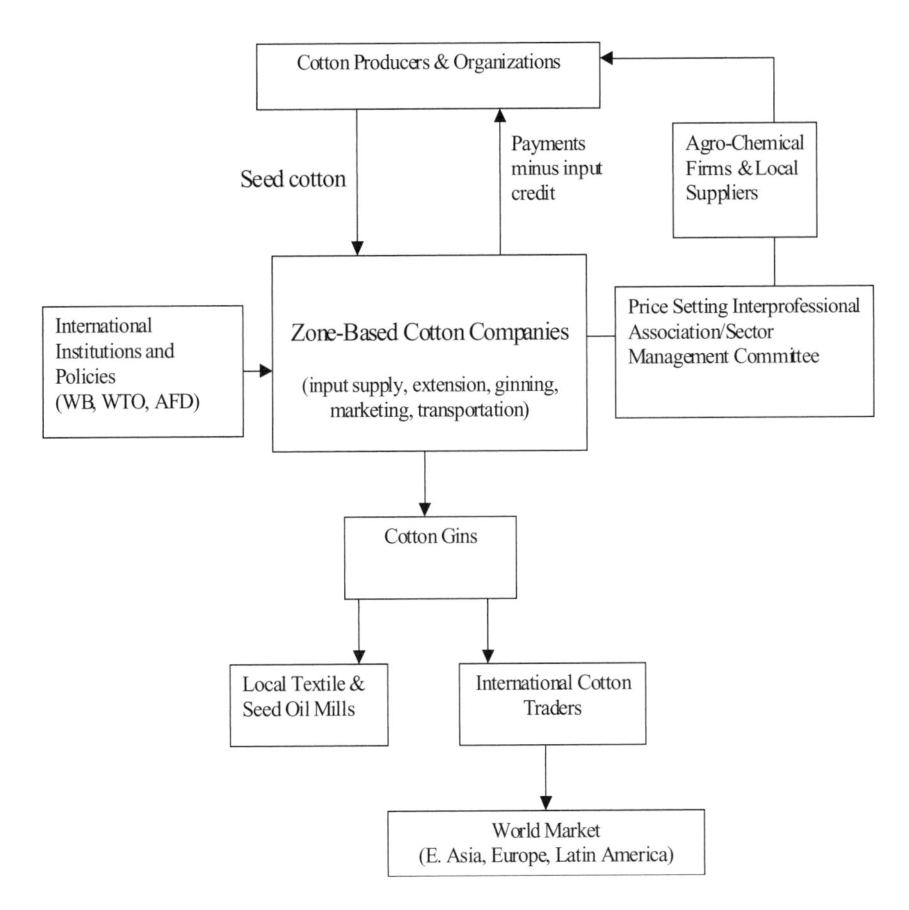

Fig. 8.3. Institutional structure of the cotton sector, Côte d'Ivoire, 2005

Under pressure from the World Bank to privatize its parastatal sector, the Côte d'Ivoire government sold six of CIDT's ten cotton mills in 1998 through an international bidding process.[3] To entice potential buyers to

[3] The two successful buyers were foreigners. *Ivoire Coton,* a company owned by International Promotion Services and Paul Reinhart, successfully bid for three gins in the Boundiali and Dianra areas. The second buyer, *La Compagnie Cotonnière Ivoirienne* (LCCI), bought three gins in the Korhogo and Ouangolodougou regions. LCCI is owned 100% by the Swiss Groupe *L'Aiglon*. Finally, the government-owned CIDT retained four gins located in the center of the country and renamed itself the New CIDT (*La Nouvelle CIDT*). The World Bank's expectation was that the New CIDT would be sold in the near future. However, the December 24, 1999 military coup and the September 19, 2002 rebellion stalled the sale of

participate in the privatization process, the state offered each winning bidder a specific cotton growing area whose inhabitants would supply it with seed cotton. Figure 8.4 shows the delimitation of these cotton zones.

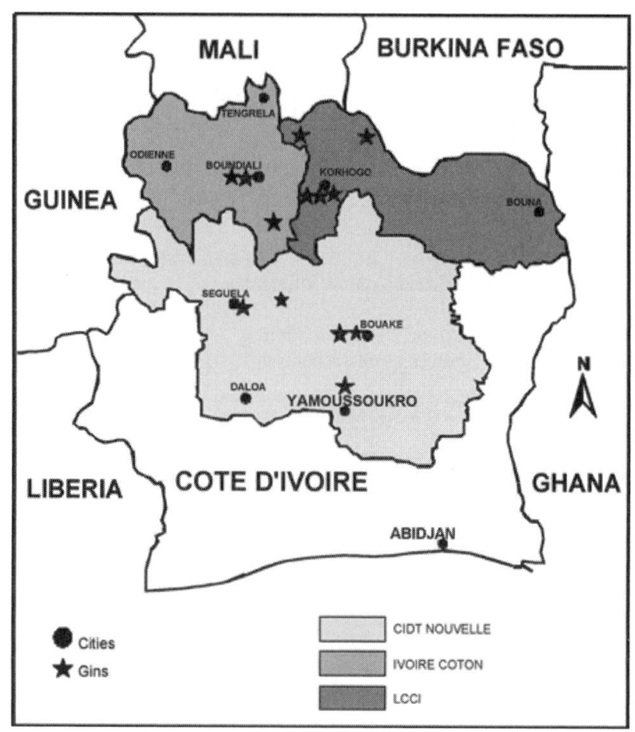

Fig. 8.4. Cotton zones of Côte d'Ivoire

Like the old CIDT, the new cotton companies were to monopolize input distribution and cotton markets in their zones. In short, the old CIDT was transformed from a national monopsony into three regional monopsonies in which CIDT continued to function as one of three companies.[4]

---

this vestige of the old order. In 2002/03, *Ivoire Coton* was the leading ginning firm, accounting for 39 percent of total production. The New CIDT and LCCI ginned 33 percent and 28 percent respectively of the cotton produced that year.

[4] CFDT contested the sale of the cotton companies in which it had a 30 percent share. It made a formal complaint to the World Bank's arbitration panel and subsequently withdrew its shares from CIDT. *La Nouvelle CIDT* is 100 percent owned by the Côte d'Ivoire state.

The extensification of cotton, 1990-2005

Farmers' commitment to cotton growing has fluctuated in response to its profitability as a cash crop within this evolving institutional framework. The two most important determinants of farmer income are the purchase price of seed cotton and the availability and cost of agricultural inputs. The two most recent declines in cotton production evident in Figure 8.2 relate to the interplay of these variables. During the late1980s and early 1990s producer prices were slashed and input costs rose as CIDT tried to reduce its budget deficits by cutting its operating costs. Producer prices for first and second grade cotton dropped 22 percent and 24 percent respectively between 1988 and 1991. Further price reductions were scheduled for 1993/94 but the government's decision in January 1994 to devalue the West African franc by 50 percent led to a 14 percent producer price increase instead. However, this reprieve was short lived. Cotton growers' costs of production dramatically increased in 1994 when CIDT ended its pesticide subsidy to cotton growers.[5] Producers now had to purchase pesticides as well as fertilizers whose basic ingredients are imported from overseas and mixed in Côte d'Ivoire. The price of agricultural inputs nearly doubled overnight as a result of the 50 percent currency devaluation.

To cut costs, farmers reduced the amount of fertilizer and pesticide applications in their fields by 57 percent. At the same time, they expanded their cotton area in an attempt to maintain past levels of production (Bassett 2001: 151-53). This extensification strategy was rational to cash poor farmers who sought to boost their incomes by reducing the costs of each kilogram of cotton produced. Although yields declined, net incomes increased. However, this strategy had its limits in light of the poor soils and input intensive cotton varieties (e.g., GL 7) that were promoted by CIDT at the time.

Official agricultural data do not capture this process of extensification. The data on the area in cotton are not reliable since cotton company extension agents no longer monitor farmers closely. As a result, they only record the area for which farmers order agricultural inputs, not the actual area under cultivation. This leads to an under-reporting of cotton area. Since cotton yields, calculated in term of kilograms of seed cotton per hectare, are derived from dividing total production by area, this under-

[5] Of course, cotton growers never received anything free from CIDT. They indirectly paid for the pesticides in the price they received for their seed cotton. CIDT set producer prices with reference to its operating expenses and world market prices. The costs of providing pesticides to cotton growers figured into CIDT expense sheet and thus led to lower producer prices.

reporting of cotton area leads to an exaggeration of cotton yields. The only production parameter that is reliable is total seed cotton production.

Cotton production increased in the second half of the 1990s and early 2000s with a record level of 402,367 tons reported in 1999/2000, and near record levels in 2001/2002 and 2002/2003. The rebellion erupted in September 2002 in the middle of the 2002/2003 growing season. Its effects on the cotton economy were more noticeable the following year when production plummeted to 170,389 tons (Eurata 2004). This figure likely exaggerates the decline in production since the rebellion severely disrupted cotton marketing and thousands of tons of seed cotton crossed the border illegally to neighboring countries (USDA 2006).

Political-economic instability at the turn of the 21st Century

The 2002 rebellion significantly disrupted the cotton economy. All of the country's cotton fields and gins are located in rebel-controlled territory. During the early stages of the conflict, rebels stole cotton-company vehicles, pillaged warehouses filled with agricultural inputs, and destroyed seed cotton stored at gins. In addition, rebels sacked the cotton experiment station in Bouaké, commercial banks closed their doors, and extension agents fled their posts. The costs of transporting cotton to port soared as truck drivers were forced to pay multiple tolls as they passed through rebel and government check points on their way to the port in Abidjan.

Poor management by the new cotton companies and a breakdown in the input supply-credit arrangement brought further chaos to the cotton sector. Between 2000 and 2004 the cotton companies LCCI and CIDT and the gin-owning farmer organization URECOS-CI, accumulated a US$60 million debt to input suppliers. They also owed farmers some $24 million for unpaid cotton deliveries over the same period (Eurata 2004, 66-67).

Escalating debts made it difficult for cotton companies to secure bank loans to pre-finance the 2005–2006 campaign. LCCI could only provide 20 percent of its farmers input needs that year (USDA 2006, 25). The combined effects of reduced inputs and the war became apparent in 2005–2006 when total production amounted to just 260,000 tons. The forecast for 2006/2007 was even worse since inputs were available for only 30 percent of the cotton growing area.[6] Just one company, *Ivoire Coton*, was paying its growers in a timely manner and recovering its input loans. On balance, the privatization of CIDT has been a failure.

---

[6] Personal communication, European Commission Delegation, Abidjan, August 31, 2006

After numerous ups and downs in the 1990s, the cotton economy was on the brink of collapse in 2005. The new companies and the institutions created to ensure their efficient operation proved incapable of regulating and coordinating the privatized sector (Europaid 2006, 87-89). Most importantly, the critical input supply-credit recovery mechanism ceased to function. More and more farmers now had to pay cash to obtain fertilizers and pesticides if they wished to grow cotton. But cash was hard to come by when cotton companies were not paying producers. The instability introduced by the war worsened these structural problems. It is no wonder that farmers were turning to alternative crops.

AGRICULTURAL DIVERSIFICATION

Disenchanted cotton growers began to invest in alternative cash crops in the mid 1990s. While the 1994 currency devaluation and the end of the pesticide subsidy worsened the terms of trade for cotton growers, the market for food crops improved. As food imports became more expensive, grain merchants looked to local producers for their supplies. The case of Katiali is illustrative. Between 1988 and 1995, cotton area declined from 47 to 34 percent of the total cultivated area. The area planted in rice and maize increased over the same time period.

Investments in tree crops

Mangoes and cashews received the most attention. Neither tree crop was new to the Katiali area. A survey conducted with 156 orchard owners in 2002 revealed that 7 percent of existing orchards had been planted in the 1960s and 1970s. A noticeable upturn occurred in the latter half of the 1980s when 15 percent of the orchards were established. However, the biggest increase in orchard area took place between 1990 and 2001. Nearly two-thirds (64%) of the orchards in existence in 2002 were created during this time.

Cashews dominate Katiali's orchards. Of the 610 hectares planted in tree crops, 85 percent were in cashews. Another 11 percent of the area included cashews inter-planted with mango, orange, and guava trees. Mango trees account for 4 percent of the total orchard area. The average orchard area was 3.9 hectares per orchard-owning household.

The agricultural geography of the tree crop boom is difficult to determine. National and regional level data on tree crop area and production are poor in quality.[7] However, we do have data on cashew exports that can serve as a proxy for production. These data show a remarkable 35-fold increase in cashew production between 1990 and 2001 (Figure 8.5). Attractive market prices provided an important stimulus. When raw cashew nut prices rose to 200 FCFA/kilogram in 1995, farmers began planting more trees (Field notes, July 9, 1995). Three years later when the trees came into production, exports soared.

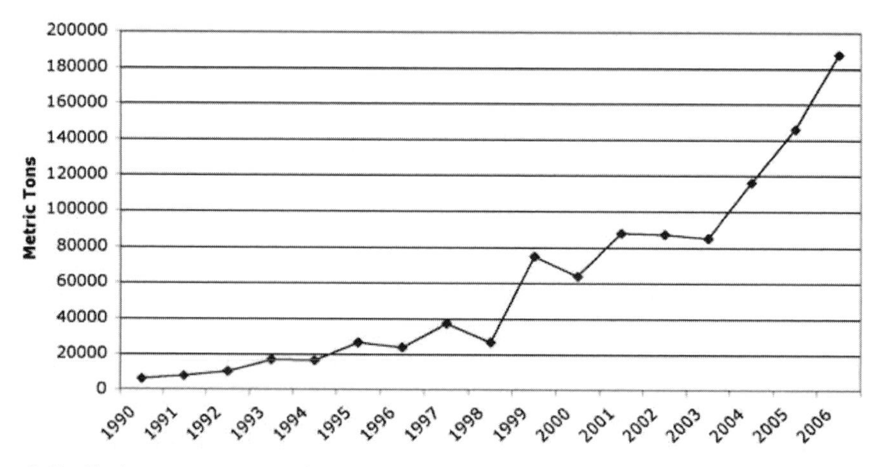

Fig. 8.5. Cashew nut exports from Côte d'Ivoire, 1990-2006

Producer prices for cashews were generally attractive over the 1990s. But the 2002 rebellion disrupted marketing and the enforcement of official market prices. In 2005 the official purchase price for cashews was 150 FCFA/kg. The price rose to 170 FCFA/kg in 2006 (Fraternité Matin, August 30, 2006). However, the weak regulation of the cashew commodity chain often results in producers receiving a lower price (Notre Voie, August 29, 2006). For example, despite the official purchase price of 170 FCFA in 2006, producers received just 125 FCFA/kg in Katiali in July of that year. Moreover, there is very little value added within the country. Just five percent of Ivorian cashew nut exports are processed within the country.[8] The bulk is exported unshelled to India (96%) and to Vietnam (4%).[9]

[7] The national extension agency, ANADER, did not have any reliable data on orchard area, production levels, and incomes when approached in August 2002.

[8] There are two cashew nut processing plants in northern Côte d'Ivoire. Cajouci is located in Korhogo and has a capacity of 1500 tons per year. It exports shelled nuts primarily to South Africa and France. The second plant, Sodiro, is located in Odienné and has a capacity of 2500 tons per year (Doumbia 2002).

TREE PLANTING AND LAND RIGHTS

The flourishing of fruit and nut orchards in northern Côte d'Ivoire is inextricably tied to individual rights to plant trees. In the Senufo culture area, individuals seeking farmland must obtain the authorization of the village chief, land priest (*tarfolo*) or lineage heads who control land passed down to them from the original village founders. Their status as customary land managers is legitimated each year when they perform ritual sacrifices to the bush spirits that inhabit the areas they manage.

In theory, Senufo land managers rarely lend land for tree planting because it would be a de facto alienation of lineage or village lands. The *tarfolo* of Katiali stated on more than one occasion that members of the Jula ethnic group, who make up more than half of the village population, would never have the right to plant an orchard. Since they are not descendants of the village founders, they are theoretically unable to acquire long-term rights to land. Yet, when confronted with evidence that some Jula households had planted orchards, the village land priest admitted that some Jula had more rights than others. Indeed, the survey of orchard owners revealed that 70 percent of the orchard area (427 ha) was owned by Jula households. Senufo households controlled 28 percent of the orchards (168 ha), and immigrants 2 percent (15 ha).

The flurry of tree planting over the past decade, especially by Jula households in Katiali, is partly explained by impending changes in how land is held in Côte d'Ivoire. A new land law passed in 1998 requires land registration and titling in the rural areas (RCI 2004). The process of registering land is a complex technical and administrative process. An important stage in the land title application process is a public hearing in the community where the land right is claimed. Members of the community have an opportunity to support or object to the land claim during the three months prior to the public hearing. It is therefore important for potential land claimants to establish their land rights well before they submit an application. The surge in orchard development can be interpreted as a tactic of local land users to stake a claim in the land that they hope will be validated during the land registration process. Because of the war, the application of the 1998 land law is currently delayed in northern Côte d'Ivoire. Once the government regains control of the area, land registration and titling will eventually take place

---

[9] The two largest exporters are Olam Ivoire, part of the Olam International Group. Olam Ivoire handled 22.1% of Côte d'Ivoire's cashew exports in 2006. The second largest international trader is APEXIM which handled 13.7 percent of cashew nut exports in 2006.

LAND LENDING ON THE EVE OF PRIVATIZATION

Even though it has yet to be implemented, the simple awareness of the new land law is already influencing land use, control, and management. This is evident in the monetization of land lending practices and the institution of grazing fees.

Land lending to land renting

We conducted a survey in 2002 among immigrant farming households in the terroir of Katiali. The survey recorded 115 agricultural camps (*vogo*) that contained some 1600 people. Questions focused on land access and modalities of payment by household. The results show that customary Senufo institutions of land lending have undergone significant changes since the early 1990s. The in-kind and symbolic gift (*tardan*) borrowers customarily give to land lenders, has given way to cash payments that can be considered rents.

Prior to the 1990s, land borrowers expressed their gratitude to customary land lenders by giving them either a token gift (e.g., a small basket of rice or peanuts) or something more substantial (e.g., a sack of rice or peanuts). In the late 1990s, around the time the land law was passed (but not applied), many land lenders explicitly required land borrowers to give them a 50 kilogram sack of rice or 10, 000 FCFA (about US$20). As of 2002/2003, this monetized payment increased by 150 percent to 25,000 FCFA.

The survey showed that the village chief, land priest (*tarfolo*), and lineage heads personally profit from land rents. In addition, certain household heads play the role of local hosts (*jatigi*) to immigrant farmers. *Jatigi* serve as intermediaries between land borrowers and customary land managers and receive both cash (*le prix de kola*) and substantial labor gifts from immigrant households for their services. Many Senufo household heads are eager to serve in this capacity to reap these benefits.

In summary, land lending is increasingly a tactic pursued by well-placed Senufo households to augment incomes that have fallen as a result of structural adjustment policies, depressed world cotton markets, and most recently, an armed rebellion. The dynamics and strategic actions surrounding land lending illuminate the positions and stakes of a range of actors in a rural world on the brink of land privatization.

Instituting grazing fees

Another recent development is the levying of grazing fees by local farmers on mobile pastoralists whose herds graze the residues of harvested fields. We tracked the movements of cattle herds owned by seven immigrant Fulani pastoralists and one prosperous Jula household over a three-year period (2002–2004). The question motivating this research is how mobile pastoralism, a highly productive form of livestock raising, will be affected by land registration and titling? This research revealed that farmers were beginning to charge grazing fees to immigrant herders before their cattle were allowed to graze crop residues and fallow fields. Local herd owners were not subject to these fees. The new practice is to charge 1500–2000 FCFA (US$3–4) per hectare depending on the field size and quantity of crop residues. One Fulani herd owner explained the negotiations involved in *buying a field*:

> *If the cotton harvest is complete and you want your herd to graze in a harvested cotton field, you have to give money first. One can no longer graze [one's cattle] for free in a field. You can buy a field for 5000 francs, 7500, 3000, 2500; you can even have some for 1000 francs. The largest fields go for 15,000.*

Fulani pastoralists try to avoid paying grazing fees by herding their cattle at night and moving on early the next day before their presence is known. Yet many appreciate the high nutritional value of crop residues for their cattle at the beginning of the dry season and are willing to pay the requisite fee.

The monetarization of grazing rights is also linked to the relative scarcity of good pasture as more land is put under cultivation and as the number of grazing animals rise. Competition for rangeland is partly linked to the ongoing immigration of additional Fulani herds to the area but also to the rise of mobile livestock raising among local farmers. Fulani herders view the expansion of cashew nut orchards as particularly worrisome. Because cashew trees are intercropped with food crops, it is difficult for cattle to graze post-harvest crop residues without damaging trees. As a result, herds have to bypass these highly valued grazing areas. One Fulani herder expressed his concerns about this land use competition and the future of mobile livestock raising:

> *It is the cashew orchards that have really been a problem for herders. All the fields have cashew nut trees planted in them. Wherever you go there are cashew trees. You will find that a single person has*

three fields and each of these fields contains cashew trees. I personally believe that if the planting of cashew nut trees does not decline, that (mobile) livestock-raising will come to an end. (Interview with Sita Sidibé, Katiali, June 30, 2006).

As a result of these changes in rangeland conditions, there is a general movement of Fulani pastoralists away from the Korhogo region towards the center west of the country around Touba. There is less farming in the Touba region and far fewer orchards. Its proximity to the international border is also attractive to the Fulani. In the event of further political instability, they will be able to cross the border quickly with their herds.

LAND COVER CHANGE

Agricultural diversification and heightened competition over land are modifying the landscape of the Katiali region. The dominant trend in land cover change has been a decline in savanna woodlands (open forests, wooded savannas, and gallery forests) and a corresponding increase in bush savanna and farmland. We compared satellite images for 1984 and 2003 of the terroir of Katiali, an area encompassing some 570 km^2.[10] Over this twenty-year period the area in woodlands declined from 35 to 13 percent of the land cover. Grass savannas also declined from 26 to 20 percent of the area. The largest increases were recorded for bush savanna and farmland (cropland and fallow fields). The area in bush savanna increased from a third (33%) to nearly one-half (48%) of the total area. Farmland also increased by 15 percent, rising from 5 to 20 percent of over this period.

[10] The remote sensing component of this research involved the analysis of two scenes acquired at different dates. The first scene, from Thematic Mapper (TM), was captured on November 25, 1984; the second image, from Enhanced Thematic Mapper (ETM+), was taken on December 7, 2003. The final image dataset for each acquisition date included TM and ETM+ bands 2 (green wavelength band), 3 (red wavelength band), and 4 (near-IR wavelength band). The unsupervised clustering algorithm (ISODATA program) was performed using Erdas Imagine on the three-band data for each of 1984 and 2003 images. 200 spectral classes for the 1984 image and 100 spectral classes for the 2003 image were reduced to a set of five thematic classes (woodland/open forest, bush savanna, grass savanna, cropland and fallow fields, and water).

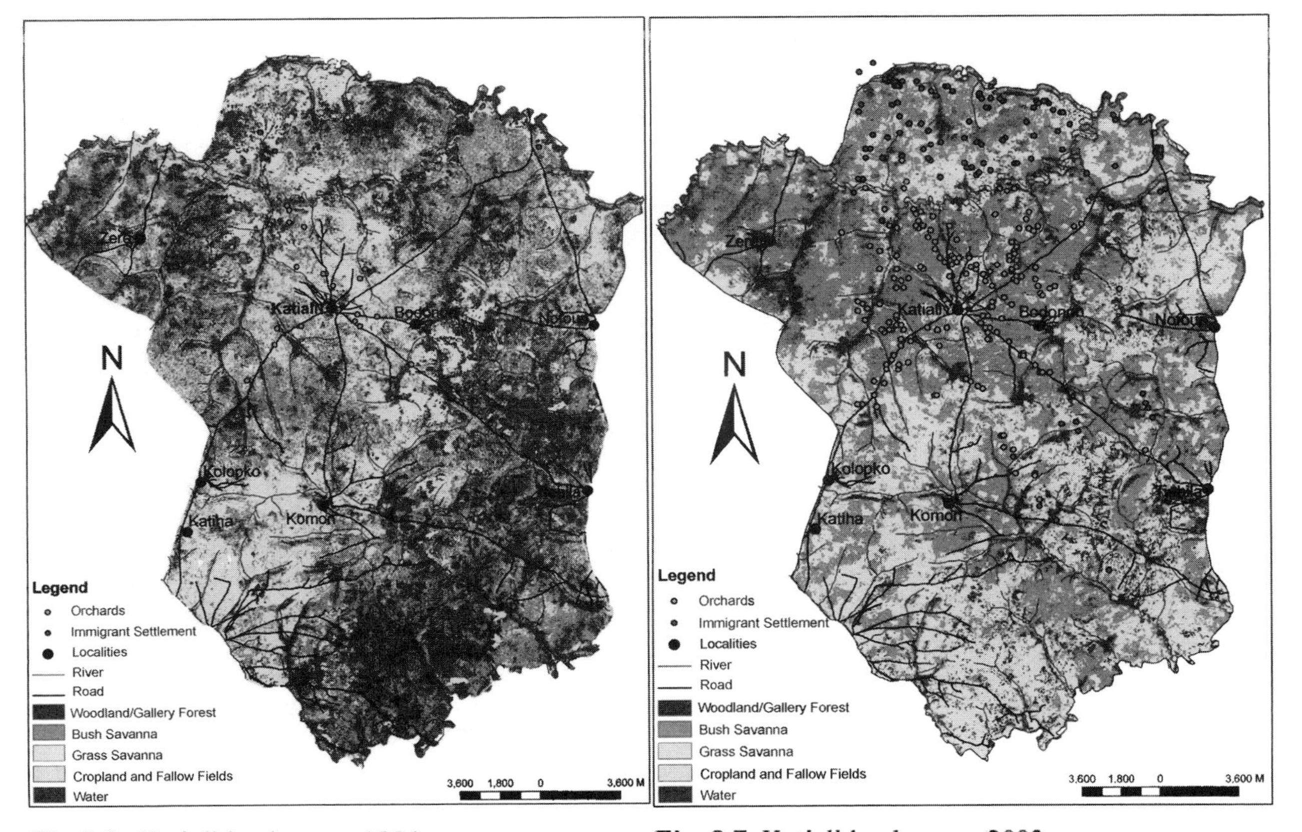

Fig 8.6. Katiali land cover, 1984

Fig. 8.7. Katiali land cover, 2003

Figures 8.6 and 8.7 reveal the geographical patterns of these dominant land-cover changes. The first pattern is the general expansion of farmland in the area. In 1984, cropland and fallow fields were concentrated in the central and southwestern regions around Katiali and to the west of Komon. By 2003, farmland had notably expanded, especially on the periphery of the *terroir*. This surging demand for land by immigrants, many of whom originate in neighboring communities, suggests a process of encroachment of farmers residing in neighboring territories. The second pattern is visible in the eastern half of the terroir where savanna woodlands and grass savanna have dramatically declined between 1984 and 2003. Land cover in this area is now a mix of farmland, bush and grass savannas, and remnant woodlands.

Figure 8.7 also records the location of immigrant settlements and orchards in the terroir of Katiali. Orchards are typically planted in fields a year or two before they are put into fallow (Figure 8.8). They are commonly found within a six kilometer radius of the village. On our satellite images, they generally appear in areas that we classify as bush savanna. The proximity of the orchards to Katiali confirms that only the Senufo and Jula of Katiali possess tree planting rights.

Fig. 8.8 Cashew trees growing in farmer's peanut field, Katiali, 2002.

The concentration of immigrant settlements in the northern half of the *terroir* is linked in part to the demand for land from people residing in the

Mbengué area to the north of the Bandama River, and partly to land disputes among land managers of Katiali. Out of the 81 immigrants who had settled in the area known as Pkala located between the Badenou and Bandama Rivers, 70 were from the neighboring town of Mbengué. These settlers were welcomed by competing land managers from Katiali who were eager to benefit from the cash and in-kind gifts demanded from immigrants.

At least four different land managers (two lineage heads, the village chief and the village land priest) claim land management authority in the Pkala area of Katiali. One lineage head claimed to control the entire area. To demonstrate his authority, he regularly granted immigrants permission to settle (Figure 8.9). He backed up his claims with recourse to ritual. He dared his rival land managers to take part in a land swearing ritual called *n'tarlé* that forbids people from cultivating an area that has a curse on it (Coulibaly 1978). Individuals who disregard this curse run the risk of personal misfortune if they farm the contested land. The curse can only be lifted by the legitimate land manager and his ritual experts known as the *n'tinbélé* who are capable of performing the *n'tarwolo* rite that lifts the curse.

Fig. 8.9 Bêh Silué, lineage head of Katiali, examining gifts offered by a young man from Mbengué seeking land to farm.

CONCLUSION

The most important dynamics driving land cover change in the Katiali region are farmer coping strategies in an ailing cotton economy and the rural land privatization law. Although the land law has yet to be implemented, it is already influencing land-lending practices and farmer investments in perennial tree crops. Worsening terms of trade for cotton growing have led farmers to engage in agricultural diversification and extensification to avoid a deterioration in living standards. The influx of immigrant farmers since the 1990s contributed to land cover changes in two ways. Immigrants cleared and cultivated land for themselves thus increasing the area in cropland. They also provided cash and labor services to land mangers and their hosts (*jatigi*) which facilitated agricultural diversification and extensification by these farmers.

Land use changes related to the expansion of tree crops and cotton are creating situations of land competition among different resource users. Mobile pastoralists complain that rangelands are diminishing with the expansion of cashew orchards. At the same time, immigrant farmers are increasingly denied their requests to put more land under cultivation because of their hosts' concerns about relative land scarcity. Requests by Jula households to expand their orchard area are also turned down by Senufo land managers. Grazing fees and land rents are now common occurrences in what is increasingly seen as a state of relative land scarcity. The heightened competition for land is characterized by disputes among Katiali's land managers over land control authority. In all cases, the imminent application of the 1998 land law is stirring individuals and groups to assert their rights in land. These assertions are manifest in the proliferation of trees and grazing fees, land renting, and competition among land mangers. These developments reveal how individuals and groups are positioning themselves in relation to the uncertainties ushered in by the new agrarian (dis)order. The case study of the Katiali region illustrates the dynamics of land use and land cover change at the local scale and how these are related to wider national and international processes linked to neoliberal economic reforms.

ACKNOWLEDGEMENTS

This research benefited from a grant made by the National Science Foundation's Geography and Regional Science Program (No. BCS00-99252).

We are also grateful to Don Luman and Koli Bi Zuéli for their assisatance with the remote sensing component of this paper.

REFERENCES

Bassett T (2001) The Peasant Cotton Revolution in West Africa, Côte d'Ivoire, 1880-1995, Cambridge University Press, Cambridge

Bassett T (2003) Le coton des paysans: Une révolution agricole. Côte d'Ivoire, 1880-1999, IRD Editions, Paris

Bassett T, Koli BZ, Ouattara T (2003) Fire in the Savanna: Environmental Change and Land Reform in Northern Côte d'Ivoire. In: T. Bassett and D. Crummey (eds) African Savannas: Global Narratives and Local Knowledge of Environmental Change in Africa, Portsmouth Heinneman and James Currey Publishers, NH and Oxford, pp. 53-71

Bernstein H (1978) Notes on capital and peasantry. Review of African Political Economy 10:60-73

Bernstein H, Woodhouse P (2001) Telling environmental change like it is? Reflections on a study in sub-Saharan Africa. Journal of Agrarian Change 1(2): 283-324

César J (1991) Etude de la production biologique des savanes de la Côte d'Ivoire et son utilization par l'homme. Thèse d'état. Université de Paris

Coulibaly S (1978) Le paysan sénoufo, Abidjan: Nouvelles Editions Africaines

Diabaté G (2002) Analyse du secteur de l'anacarde. Situation actuelle et perspective de développement, Centre de commerce international CNUCD/OMC (CCI) http://www.unctad.org/infocomm/francais/anacarde/Doc/ivoire.pdf

Dibi P (2004) Rainfall and Agriculture in Central West Africa since 1930. Ph.D. thesis, University of Oklahoma

Eurata (2004) Côte d'Ivoire: Etude des measures d'urgence pour l'amélioration de la performance de la filière coton. Rapport Final, Commission Européenne

Europaid (2006) Côte d'Ivoire: Elaboration d'une stratégie sectorielle coton: Perspectives à moyen et long termes: Rapport provisoire de première étape. Diagnostic et propositions d'orientation. Août 2006, Commission Européenne, Brussels

Fichet M (1997) Côte d'Ivoire: Derrière le concept de privatization. Le démantèlement d'une filière, Coton et Développement 22:2-3

Fraternité Matin (2006) Anacarde: Le prix d'achat fixé à 170 francs, Fraternité Matin, August 30, 2006

Goldman M (2005) Imperial Nature: The World Bank and Struggles for Social Justice in the Age of Globalization .Yale University Press, New Haven

Notre Voie (2006) Anacarde—300 participants réfléchissent sur la redynamisation de la filière. Notre Voie, August 29, 2006

Tsing A (2005) Friction: An ethnography of global connection. Princeton University Press, Princeton

République de Côte d'Ivoire (RCI) (2004) Projet Nationial de Gestion des Terroirs et d'Equipement Rural (PNGTER) Recueil des Textes Relatifs à la Loi sur le Domaine Foncier Rural, Abidjan

United States Department of Agriculture (USDA) (2006) Cotton and Products: West African Region. Benin, Burkina Faso, Côte d'Ivoire, and Mali. GAIN Report IV5010

CHAPTER 9 - Village Settlement, Deforestation, and the Expansion of Agriculture in a Frontier Region: Nang Rong, Thailand

Barbara Entwisle[1], Jeffrey Edmeades[2], George Malanson[3], Chai Podhisita[4], Pramote Prasartkul[4], Ronald R. Rindfuss[1], Stephen J. Walsh[5]

[1]*Department of Sociology & Carolina Population Center, University of North Carolina, Chapel Hill, NC 27599-3210, USA*
[2] *International Center for Research on Women, Washington, DC 20036, USA*
[3]*Department of Geography, The University of Iowa, Iowa City, IA 52242, USA*
[4]*Institute for Population and Social Research, Mahidol University, Nakhorn Pathom 73170, Thailand*
[5]*Department of Geography & Carolina Population Center, University of North Carolina, NC 27599-3220, USA*

ABSTRACT

The settlement of a frontier is the outcome of demographic processes occurring within a changing socio-spatial context and subject to biophysical constraints. This chapter elaborates on these interrelationships using the history of the last 50 years in Nang Rong, Thailand. It documents two eras in village settlement, a period of colonization prevailing in the 1950s and 1960s when entirely new villages were being formed, and a period of consolidation in the 1970s, 1980s, and 1990s that consisted of the expansion of existing villages. Spatial and temporal patterns of settlement reveal clues about the factors influencing village settlement, including the importance of suitable land for cultivation, proximity to water, markets and prices, road access, and social factors linked to proximity to other villages. The impact of initial and expanded settlement on deforestation and conversion of land to agricultural uses over four decades is examined using aerial photography.

INTRODUCTION

Frontier settlement and the rapid conversion of forests to agricultural uses is one of the more visually dramatic examples of the connection between population and environment. Although settlement is sometimes planned and directed, where people settle is mostly a choice. Other things being equal, new settlers will prefer relatively flat, high-quality land proximate to water sources that offers security, sociability, and easy access to markets. New settlers will tend to settle first in such places, and over time, may settle more densely there than other places. Features of the land and the timing and density of settlement, in turn, affect decisions about land use within a context of available technology, markets, and institutional arrangements. When new markets develop, policies change, or new technology is introduced, already settled places can be redefined as new frontiers. We elaborate on these interrelationships based on the history of village settlement in Nang Rong, Thailand (Figure 9.1).

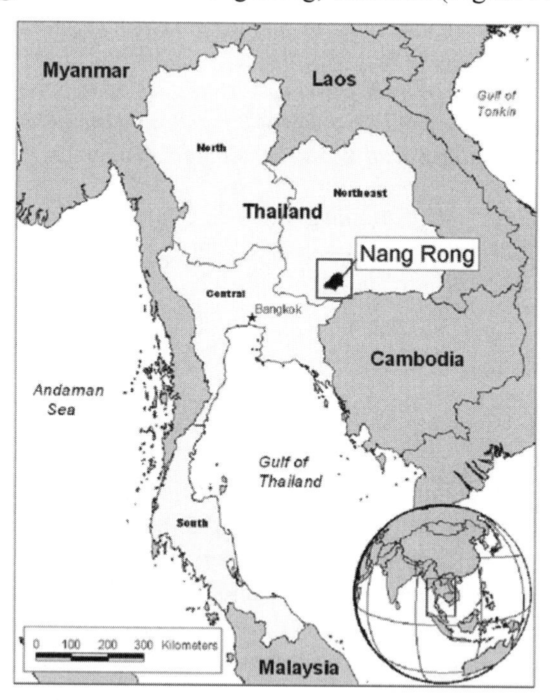

Fig. 9.1. Nang Rong District, Thailand

VILLAGE SETTLEMENT PROCESSES

Settlement is a dynamic process. Given the availability of new lands to cultivate in a frontier area, timing and spatial patterning will be affected by biophysical, economic, and social factors operating at several levels. In Nang Rong, the environmental setting is one of marginality, relatively low soil fertility, unpredictable precipitation, insufficient drainage, and generally speaking, a limited natural resource base. Biophysical factors relevant to settlement processes are characteristics of the land (slope, elevation, vulnerability to flooding, soil characteristics, and suitability for cultivation) and accessibility to water. Social and economic factors include size of the settlement, proximity to other villages, and proximity to transport routes as well as land tenure arrangements, market conditions, and government policies (or the absence thereof). The impact of settlement on land use and land cover will depend on timing as well as these other aspects of context. Figure 9.2 shows our conceptual model.

In 1950, Nang Rong supported a subsistence economy based on rainfed paddy rice cultivation. New settlers would have been looking for the best available land for rice cultivation in combination with access to drinking water and a place where houses could be built and safely maintained. Basic soil characteristics, relative elevation, slope, and accessibility to water are the main biophysical characteristics that determine suitability for rice cultivation in this context (Phongphit and Hewison 2001). Swampy low-lying ground around rivers is particularly good for this purpose (Riethmuller et al. 1984: 154). Village residents would want to live close to their fields (farm implements had to be carried back and forth to the village, as did the rice when it was harvested), close to a water source for drinking water for themselves and their animals (Phongphit and Hewison 2001), and ideally, on a small hill or rise rather than in a low-lying area (Riethmuller et al. 1984).

Features of the land that is good for rice cultivation are not necessarily good for the cluster of dwelling units. Low-lying flat land close to a river might be excellent for paddy rice cultivation but because of flooding, might not be a good place to build houses. We have anecdotal evidence (although no hard data) that some groups of people initially settled in flood prone areas which, presumably, were also good rice growing areas. Then after one, two, or more severe floods they abandoned settlement in that specific spot on the landscape. We expect that the likely scenario was that they then moved to nearby higher ground. Hence, a few villages probably experienced migration of dwelling units while retaining use of the same productive rice land.

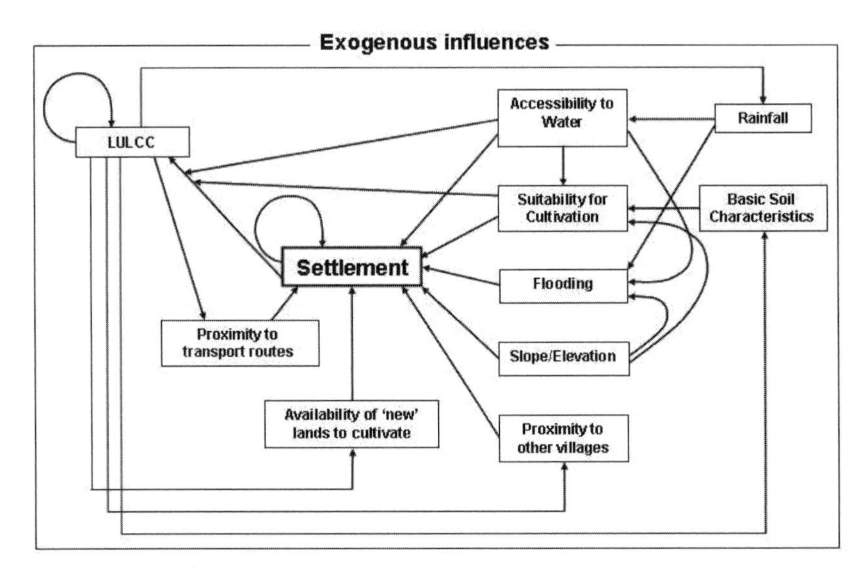

Fig. 9.2. Conceptual model of village settlement pattern

Settlement is also affected by proximity to other villages and transportation routes. There are several possibilities. People might prefer to settle relatively close to already established villages to take advantage of potential economies of scale and infrastructure that might have already been developed (e.g., roads, schools, temples) as well as for sociability and potential help in a time of need (a crude form of insurance). Alternatively, if new settlers wanted a place to maximize expansion in the future, new villages might appear in places good for cultivation and somewhat distant from other villages existing at the time (*cf.* Hudson 1967, 1969).

After a new settlement is established, land cover around the settlement changes. A village cluster would be composed of houses, vegetable gardens, and public areas. Connector paths and roads would be built linking the settlement to other villages and to a main road, if one exists. Land around the village cluster would be converted to cultivation in a pattern radiating out from the village but influenced by biophysical characteristics such as topography, the spatial distribution of soils, presence of streams and rivers, and the like.

These changes in land use would affect subsequent settlement, both new settlement (the creation of a new village nearby) and expanded settlement. Once a new village is established, it grows and changes for many of the same reasons that people initially settle there.

Given a subsistence rice economy, new and expanded settlement would continue until all of the land suitable for rice cultivation was claimed. A time would come when the frontier closed a threshold consequent to feedbacks built into the system. In the case of Nang Rong, however, two exogenous changes occurred that affected relationships and feedbacks within the model: (1) the extension of the national highway system in the 1950s and 1960s, and (2) the emergence of an international market for cassava in the 1960s. These were watershed events, providing new opportunities, redefining the frontier, and stimulating further settlement.

Nang Rong is located along the Chokechai-Dej Udom Highway, which we will call the East-West Highway. This highway is part of a larger national system that was the focus of development in the 1950s and 1960s (Ingram 1971; Siamwalla 1995; Van Landingham and Hirschman 2001). It connected Nang Rong to markets in Nakhornratchisima (Korat), a major center for the processing of cassava, at that time a new crop. It facilitated the movement of heavy tractors into the district, especially parts in the southwest, to assist with the cutting of trees, plowing under of grasses, and the preparation of plots for cultivation of cassava and other upland crops. It also facilitated the movement of settlers to the region. The East-West Highway changed the proximity of villages to major routes of transportation, changed the desirability of different locations, and probably also changed the effect of settlement on land use. National Highway Department archives indicate that this highway was planned in the 1940s, a dirt road in the 1950s, and paved during the mid to late 1960s (probably accelerated to serve military needs related to the Vietnam War and the communist insurgency).

The emergence of an international market for cassava was also critical to settlement processes and their effects. Demand for this new cash crop originated in European Economic Community (EEC) countries in the early 1960s, where it was used as an animal-feed supplement (Riethmuller et al. 1984; Snyder 1993). In Nang Rong, land not suitable for paddy rice cultivation could acquire commercial value for cassava cultivation. As agriculture shifted from subsistence to commercial agriculture, proximity to roads would increase in importance for village settlement and change its effects on land use. Accessibility to water, as an indicator of soil suitability, would likely lessen in importance (although ready access to drinking water would still be a crucial consideration). The East-West Highway was under construction and completed at the same time as a market for Thai cassava appeared in the EEC in the 1960s.

PATTERNS OF VILLAGE SETTLEMENT

In Nang Rong, villages are clusters of dwelling units surrounded by agricultural lands. Security (safety in numbers), power (locally), and returns to scale through cooperative labor are all enhanced by this residential pattern (Phongphit and Hewison 2001; Riethmuller et al. 1984). When clusters of households reach a certain size (typically 100, although it varies), they are designated an administrative village. When the number of households grows too large (typically over 200), the village is divided administratively into two or more villages. Growth of the population thus is reflected in increases in the number of administrative villages.

Over the period of interest, there are two processes by which settlement spread over the Nang Rong landscape: new settlement, i.e., the creation of entirely new villages by settlers moving into the area; and expanded settlement, i.e., the growth of pre-existing villages through in-migration and/or natural increase. For our purposes, a new or original village is defined as one having no administrative linkage with any other village at the time of its creation. We look at both the timing of establishment and spatial distribution of original villages as well as the timing and spatial distribution of increases in the size of these villages, viewing administrative splits as an indication of village growth. The former represents colonization and the latter represents spread (Hudson 1967, 1969).

We use three pieces of information collected in the 2000 Nang Rong Community Survey[1] to track the history of villages: the year that the first village headman was appointed; whether before that, the village had been administratively part of another village; and if so, which other village. New settlement is revealed by the establishment of original villages, i.e., ones which at their founding, had no prior administrative link with any other village.[2]

---

[1] http://www.cpc.unc.edu/projects/nangrong/2000/comm_quest00.pdf

[2] This approach will miss villages that were disbanded for some reason. Anecdotally, we know of two villages that moved, in one case because of flooding problems and in

Subsequent splits indicate expanded settlement, assuming that it is growth in the number of households that leads to the split. It is not possible to examine this assumption for early decades in the period of interest in this paper, regrettably, but we can examine it in the latter part of the period. The first wave of the Nang Rong Household and Community Surveys were fielded in 51 villages in 1984. Between 1984 and 1994, 20 of those villages split administratively (a few more than once). Larger villages (measured in terms of number of households) were more likely to subdivide than smaller ones. Each household in 1984 increased the likelihood of a split by four percent, a fairly large effect. These and other similar results for the 1984-2000 period suggest that it is reasonable to interpret administrative subdivision as a measure of population growth, especially growth in the number of households.

Of the 352 villages existing in 2000, 269 (76 percent) were established in or after 1950, the starting point for our analysis. Figure 9.3 shows trends in the numbers of original and administrative villages. In 1950, there were 61 original villages, a number which increased to 74 by 1970, 77 by 1975, and 82 by 1990. After this time, no completely new villages were established. In 1950, there were 80 administrative villages. This number grew to 142 in 1970 and to 352 in 2000. Putting the two trends together indicates that over the last 50 years, there were two eras in village settlement, one prevailing in the 1950s and 1960s when entirely new villages were still being formed, and the other in the 1970s, 1980s, and 1990s that consisted mainly of expanded settlement. The shift from one era to the next coincides with the completion of the East-West Highway and the opening of the cassava market.

Using locational coordinates for villages collected in 2000 and digitized information about hydrography and the road system organized within a GIS, three sets of proximity measures were calculated: distance to the nearest village; distance to the nearest water source; and distance to the major East-West Highway. Distance was calculated as Euclidean distance.[3] Trends in proximity to the nearest village for quinquennial time points between 1950 and 2000 are shown in Figure 9.4 for original villages and administrative villages.

---

another case to move closer to a main road (also see Siamwalla 1995). We have not heard of any villages that were disbanded entirely.

[3] We did make one adjustment, though. When the nearest village according to Euclidean distance is across a river, because the river would impede social contact, we turn our attention to the next nearest village (as long as it too is not across a river). Likewise, we avoided rivers in calculating distances to the East-West Highway.

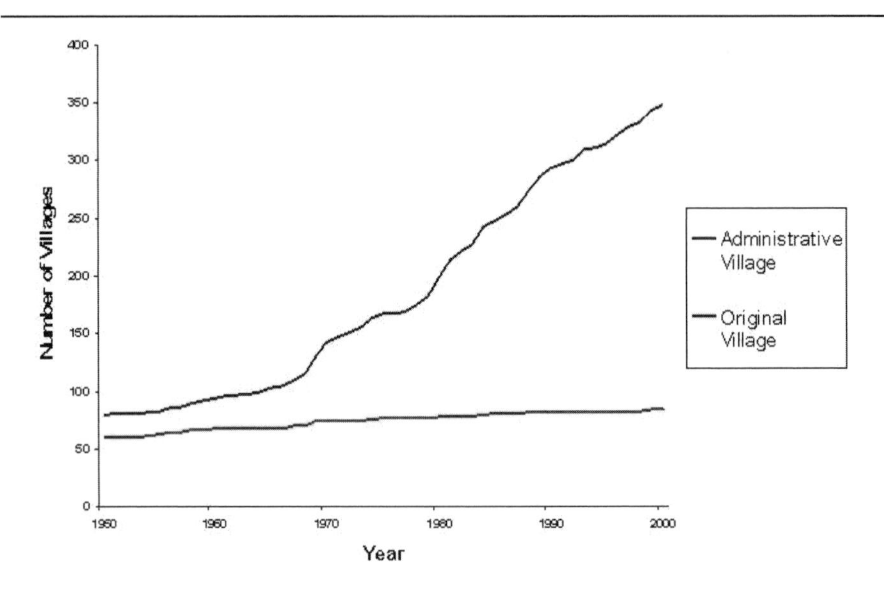

Fig. 9.3. Number of original and administrative villages, 1950-2000

There are contrasting expectations about trends in distance to the nearest village. On the one hand, to take advantage of sociability, mutual aid, and emerging infrastructure, new villages might locate close to existing ones, other things being equal. Distance to the nearest other village will not change much through the settlement process, even though the total number of villages may be increasing. On the other hand, if new villages are established within the matrix of villages already present, as predicted by Hudson's dynamic theory of central place settlement, distance to the nearest village would decrease over time. In 1950, the average original village was a little over two kilometers from the nearest other original village. It was a little closer to the nearest village in 2000, but the difference is slight (1.9 rather than 2.1 kilometers). As shown in Figure 9.4, there is a slight downward trajectory in the cumulative average, but overall the trend is fairly flat, suggesting the importance of social in addition to ecological factors as determinants of village location.

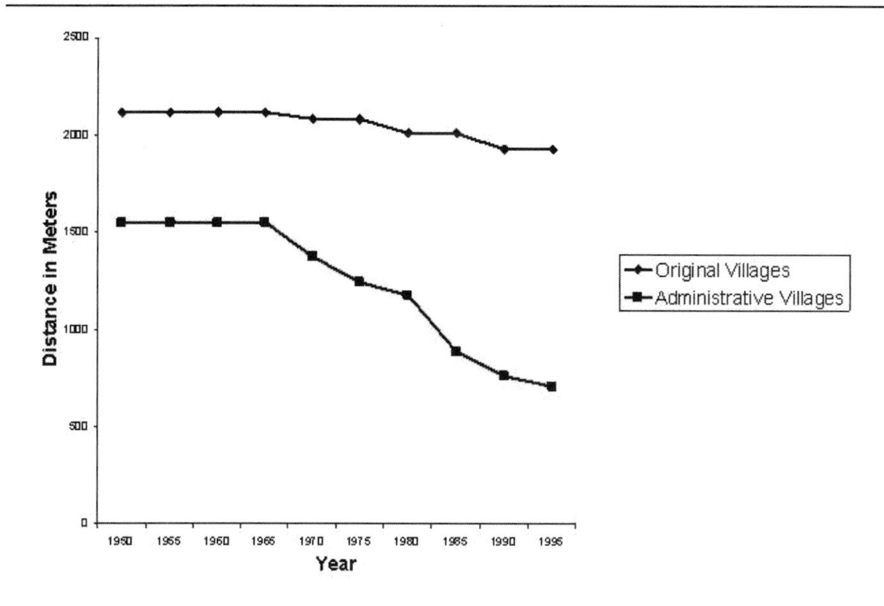

Fig. 9.4. Distance to nearest village

This pattern stands in sharp contrast with that for administrative villages. The average distance to the nearest administrative village did not change between 1950 and 1970, but after that it took a dramatic turn downwards. In 2000, as a consequence of expanded settlement, administrative villages were half again closer to the nearest administrative village (0.6 km versus 1.3 km). This shows the consequences of a shift in patterns of village settlement, from colonization in the 1950s and 1960s to village expansion thereafter.

Turning now to proximity to a river or stream, if settlers are reliant on paddy rice cultivation, they will locate villages where average distance to the nearest significant water body is relatively short. This average may lengthen somewhat over the settlement process as the early settlers select the best locations, leaving less desirable locations for those arriving later. To the extent that settlers grow other crops such as cassava, villages can be sited where distance to water may be a little longer.The development of an international market for cassava as well as the roads to connect Nang Rong to that market could have changed the relative importance of water accessibility, with consequences for the timing and spatial patterning of settlement. If so, then it should be possible to document a shift in proximity to a water source after 1970. In fact, the trend for original villages is flat. Median distance to a water source did not change much over the 50 years of observation (rounding to 0.5 km for each date in the series). For administrative villages, in contrast, median distance to a water source grew steadily through the period, with increases particularly marked between 1955 and 1975. The contrast between the trends for original villages and administrative villages suggests that the introduction of market agriculture encouraged expanded settlement, especially into areas farther away from rivers and streams. With respect to water for domestic use, underground springs as well as streams and rivers can serve this purpose.[4]

Finally, we consider proximity to the East-West Highway, completed in the late 1960s or early 1970s. The route followed by this highway was determined by distant engineers and planners, not by a political process in which local villages competed to have the new road pass through their territory. In other words, the location of this highway is exogenous to the processes of village settlement under study. Because it connected the district to markets for upland crops, notably cassava, we thought that new villages would locate close to it, and already established villages that were close would be more likely to grow and expand than those farther away from it. These expectations were not met. The average distance of original villages from the (future) road grew rather than decreased over the period, from 5.9 kilometers in 1950 to 6.4 kilometers in 1970 and 6.9 kilometers in 1990.In other words, new settlements were located increasingly far away from the road, not closer to it. Moreover, not only were new villages established during the period likely to be sited in areas somewhat distant from the (future) highway, village expansion was also more pronounced in those areas. Between 1950 and 1970, distance from the average administrative village to the East-West Highway increased from 5.8 to 7.3 kilometers. Distance increased especially rapidly between 1965 and 1975, when the highway was constructed and also when international demand for cassava

[4] Qualitative data gathered in 2003 suggests that the presence of underground springs affected the location of several villages in the southwest Discussions with older residents of these villages made it very clear that the presence of underground water near the surface led to the founding of villages along this arc. When looking for possible sites to start a village, information about underground springs was obtained by potential settlers from a number of sources. First, the vegetation was distinctively different when water was close to the surface. Second, large mammals would scratch the ground to expose the water for their own drinking. Hunters would discover these places and, in turn, would tell those looking for a village site. There was some timbering occurring in the area, and those looking for an appropriate village site would also learn about sub-surface springs from loggers.

exploded. With benefit of hindsight, we suspect that feeder roads were constructed at the same time or soon after, opening up interior areas, especially in the southwest.

SETTLEMENT AND LAND USE

Aerial photos for 1954, 1966/67/68, 1983/84/85, and 1994 were scanned, registered, rectified and mosaicked together to track changes in land cover in relation to new and expanded village settlement. To illustrate these changes, we selected two sites, one in a low-lying area suitable for paddy rice cultivation and the other an upland site more suitable for cassava cultivation. Villages were established after 1954 in both sites. The village in the lowland site was established in 1984, which subdivided twice subsequently. There are two villages in the upland site, one established in 1969 which split in 1972, 1988 and 1999, and the other also established in 1969, but which split in 1979, 1982, 1987, 1988, 1998 and 2000.

The location of each site, defined by a 3 x 3 km window, is shown in Figure 9.5. Within the box, for each of the four aerial photos, land cover was visually examined and classified into six categories: forest, rice, upland agriculture (i.e., cassava, sugar cane, maize, *kenaf*), scrub, water and urban (i.e., built up) areas. We track the percent of the land cover in forest, rice, upland agriculture, scrub and urban uses over the four images. We also track trends in the fragmentation of the landscape, focusing on the number of patches.

In 1954, the upland site was almost completely covered in forest. Change over the next four decades was dramatic. Forest cover declined from 98 percent in 1954 to 13 percent in 1994, with the decline particularly marked in the 1967-1984 period. Forest was replaced by upland agriculture. Conversion was underway by 1967, by which time 15 percent of the area was in upland crops, but more than tripled over the 1967-1984 period to 65 percent. The two villages in this site were formally established at the beginning of the period (1969) and grew rapidly, as measured by administrative subdivisions. Change in land cover both anticipated and followed the completion of the East-West Highway, but was more pronounced afterwards.

Deforestation and the expansion of rice production in the lowland site proceeded at a more leisurely pace than the expansion of cassava production in the upland site, starting earlier and continuing longer. In contrast to the upland site, which was almost completely in forest at the beginning of the period, conversion to agriculture was already underway in the lowland site at the beginning of the observation period. In 1954, 23 percent of the area was in paddy rice cultivation. The fraction grew slowly to 30 percent in 1967, almost doubled in the next 15 years to 57 percent, and continued to grow to 72 percent in 1994. Forest cover declined correspondingly, from 75 percent in 1954 to 22 percent in 1994.

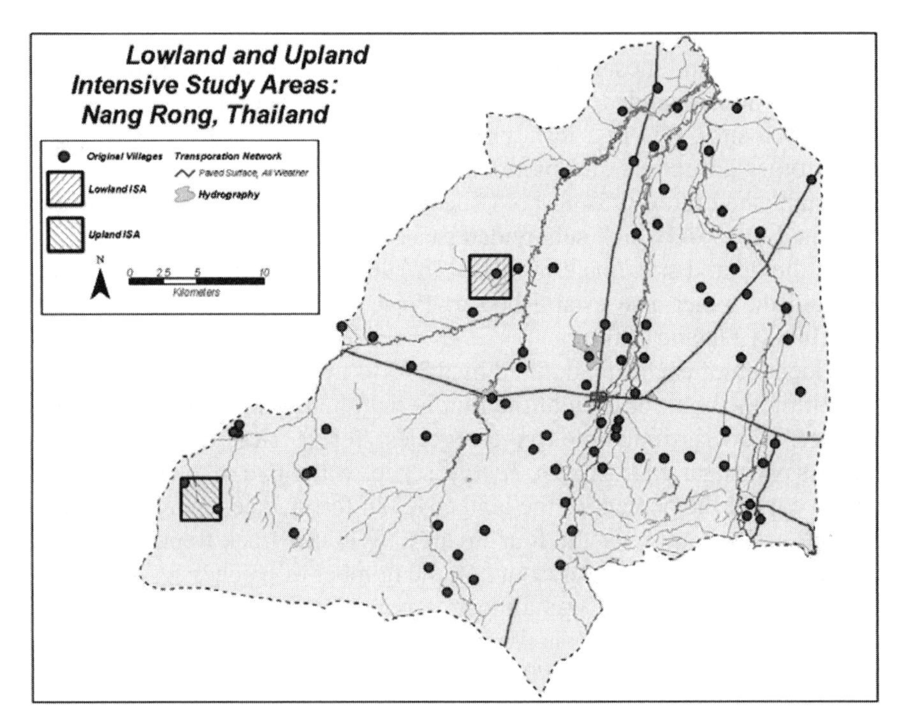

Fig. 9.5. Location of lowland and upland intensive study areas.

In both sites deforestation and conversion to agriculture are associated with changes in the spatial organization of the landscape. Forests were fragmented. The number of forest patches increased from 21 in 1954 to 42 in 1994 in the lowland site and from 7 to 104 over the same period in the upland site. Location also changed over this period. When settlers moved into the area, and new villages were established, forest was located in the periphery, away from village settlements. By 1994, this pattern was reversed. Outlying forest had been converted to agriculture. Forest patches tended to be small, located in and around the village settlements themselves.

Whereas forests were fragmented, there was consolidation in the land devoted to upland crops and especially rice cultivation, over time. In the lowland site, the number of rice patches decreased from 134 in 1954 to three in 1994. Initially, given the work involved in the creation of paddy and the building of bunds, and later from the standpoint of cooperation and water control, it is advantageous to locate rice fields contiguously. In the upland site, the number of patches in upland crops decreased from 83 in 1967 to 47 in 1994. Contiguity is not quite so important in the siting of the fields for upland crops. There may also be more local variability in topography in the upland than lowland regions.

CONCLUSION

An exploration of village history in Nang Rong, Thailand and trends in the spatial distribution of villages and the composition and spatial organization of the landscape at two sites shed light on our model of village settlement (Figure 9.2). Specifically, it points to the importance of new and expanded settlement and land characteristics within a changing historical context. In the early years, when subsistence agriculture prevailed in Nang Rong, suitability for rice cultivation was key to village settlement. Areas that were suitable for rice were settled first. Entirely new villages were created. The impact of human activity on the landscape was visible before villages were established, but accelerated after that. Land use change appears to have radiated outward, a spatially cumulating process following local topography and recognizing the presence and location of other nearby villages. In the mid to late 1960s, as a result of several exogenous changes, market agriculture was introduced into the district. The building and paving of the East-West Highway linked Nang Rong to growing national and international markets for agricultural products, and also allowed for tractors to come in. In the lowlands, expanded settlement increased the extensification of rice production. In the uplands, it made marginal lands economically viable. Indeed, expanded village settlement in the upland areas was associated with rapid conversion to upland crops. Conversion was not as homogenous spatially as in the lowlands, but the basic pattern was the same.

Time is an important element in our vision of how villages are settled and their resulting spatial patterns. Historical migration waves, changing rates of natural increase, the timing of village splits, and the change in the densification of settlement patterns are all time dependent factors. Accessibility changes as new roads are built, existing roads improved, and connections between villages are altered through village formation. Exogenous factors may be episodic with associated time lags, or events may happen at a discrete point in time thereby altering subsequent events. Monsoonal rain patterns and periods of drought are examples of the former. The sudden upturn in demand for cassava in Europe in the late 1960s is an example of the latter. There are many ways in which time is reflected in village settlement patterns and in patterns of land use, which may have interesting feedbacks to village settlement. While already suggested above, we emphasize some of the more important time-dependent elements here to underscore their importance in creating possible triggers of action, altering the initial conditions for village formation or evolution, and setting and resetting the dynamic equilibrium of the system.

ACKNOWLEDGEMENTS

We would like to thank NASA (NAG5-11290) and NSF (SES0119572) for funding this research. We would also like to acknowledge NICHD (R01-HD25482) for supporting the collection of data on which it is based. We greatly appreciate assistance with the data received from Philip McDaniel and Richard J. O'Hara, and help with graphics and manuscript preparation received from Tom Swasey and Bridget Riordan

REFERENCES

Hudson JC (1967) Theoretical Settlement Geography. Ph.D. Thesis, Department of Geography, University of Iowa

--------- (1969) Location Theory for Rural Settlement. Annals of the Association of American Geographers 59:365-381

Ingram J (1971) Economic Change in Thailand 1850-1970. Stanford University Press, Stanford

Phongphit S, and K. Hewison (2001) Village Life: Culture and Transition in Thailand's Northeast. White Lotus, Bangkok

Riethmuller R, Scholz U, Sirisambhand N, Spaeth A (1984) Spontaneous Land Clearing in Thailand: The Khorat Escarpment and the Chonburi Hinterland. In: Harald Uhlig (ed) Spontaneous and Planned Settlement in Southeast Asia. Institute of Asian Affairs, Hamburg, pp 119-248

Siamwalla A (1995) Four Episodes of Economic Reform in Thailand 1958-1992. Thailand Development Research Institute, Bangkok.

Snyder F (1993) European Community Law and International Economic Relations: The Saga of Thai Manioc. European University Institute, San Domenico

Van Landingham M and C Hirschman (2001) Population Pressure and Fertility in Pre-Transition Thailand. Population Studies 55:233-248

CHAPTER 10 - Market Integration and Market Realities on the Mexican Frontier: The Case of Calakmul, Campeche, Mexico

Eric Keys

Department of Geography, University of Florida-Gainesville, FL 32611 USA

ABSTRACT

Agricultural change, primarily the move to a hybrid of market and subsistence crops, can lead to land sparing or land expansion. In southern Mexico's Calakmul Municipality the mix of commercial chili cultivation and subsistence maize agriculture results in significant increases in land-cover change from forest to non-forest on the edges of a large conservation area. While commercial agriculture can provide farmers with higher incomes on their land, market volatility encourages farmers to grow and diversify their agricultural portfolios with some of the profits from chili sale. Market intermediaries and chili pests and plagues further confound price insecurity for chili, further pushing farmers to adopt multiple land uses. The land sparing promise of commercial agriculture in the developing world is challenged.

INTRODUCTION

Within the last 50 years commercial agriculture has spread throughout the tropics, frequently in forested areas that are valued as conservation targets (Geist and Lambin 2001; Rudel 2004; Keys and McConnell 2005). Commercial agriculture represents a shift from subsistence agricultural activities and usually is associated with more intensive labor, mechanical and chemical inputs. This increased input intensity corresponds with an anticipated increase in outputs, either in the yield of the crop or in terms of the cash earned from commodity cropping (Turner and Brush 1987; Netting 1993). While many studies note technical practices (e.g., Wilken 1987; Turner and Brush 1987) associated with commercial agriculture, few attempt to understand how it develops and how farm-to-market structures influence the amount of land used by farmers in tropical, developing regions (Watts and Goodman 1997; Whatmore and Thorne 1997; Gereffi and Korzeniewicz 1994).

I argue that commercial agriculture in southeastern Mexico's Calakmul Municipality, Campeche State (Figure 10.1), results in greater land-cover conversion than subsistence agriculture due to factors related to the marketing arrangements for the primary crop there, the jalapeño pepper (*Capsicum annuum* L., henceforth chili). Calakmul is a designated hotspot of deforestation and the focus of numerous non-governmental and governmental organizations efforts designed to curb environmental change (Boege 1995; Galletti 1998; Achard et al. 1998). Forest conversion rates in Calakmul outpace other areas in Mexico and match or exceed regional averages (Cortina Villar et al. 1999). Contrary to hopes raised by intensive agriculture for conservation's sake (Balmford et al. 2005; Green et al. 2005), commercial farmers in Calakmul confront an entrenched marketing structure that awards them meager profits and presents a perceived and real threat of market price failure in any year (Reyes 1999; Keys 2005; Keys and Roy Chowdhury 2006). Experiences with price failures for chili encourages farmers to expand the cultivation of all of their crops – subsistence as well as commercial – to buffer the blow from a potentially catastrophic market price.

The origin of chili cultivation in Calakmul can be traced to three farmers who introduced the crop in 1985. By the 1999–2000 growing season, 85 percent of farmers in southern Calakmul farmed chili for the market and were tied to market intermediaries, locally known as *coyotes* (Keys 2005). Coyotes maintain an oligopsony (a condition in which a handful of buyers control the purchase of a commodity) on chili peppers and dominate contact with national-level chili buying companies. Chili farming in Calakmul is part of a network involving smallholder producers, several critical agents among them, middlemen who control the flow of chili to the national market, the chili companies, and, of course, the market system of Mexico. In Calakmul, farmers exert little control over the chili-marketing network. This network existed long before it was imported into Calakmul and exhibits a path-dependent character. History and a national scope, in this case, play an important role in explaining the current situation of chili farmers in Calakmul.

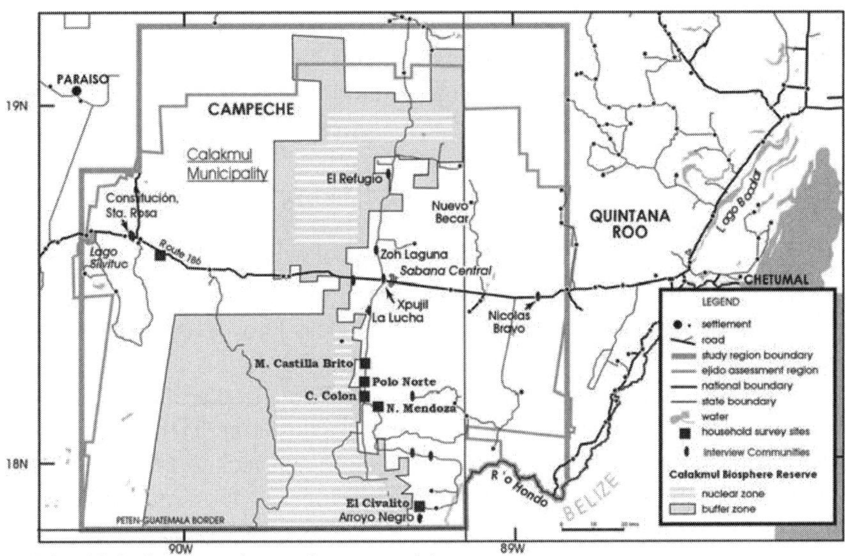

Fig. 10.1. Study region and communities

This chapter proceeds with a short discussion of prior research and theories on agricultural development. This section is followed by a description of the study area and a discussion of methods used in this study. This descriptive section contains narrative information on the origin and spread of chili cultivation in Calakmul. Following this I present descriptive statistics on the land-use portfolios of different farmer types. In the subsequent section I discuss the drivers of the witnessed land use portfolios. I end the chapter by exploring ways that marketing structures influence land use and land cover in the developing world.

AGRICULTURAL DEVELOPMENT AND SMALLHOLDERS

Agricultural changes take different, purposeful forms. "Ecosystem modification in agriculture is not random; it is directed toward the production of goods that have value, whether consumed by the farmer or sold" (Wilken 1987). Given initial success, or the promise of future success, farmers expand their cultivation through a process known as extensification; that is increasing the area under cultivation while maintaining a relatively low labor-to-output ratio (Barlett 1976; Turner and Brush 1987). In frontier areas abundant in land, such as Calakmul, farmers frequently opt to extensify so that they can maximize the returns to labor and capital investments (Schmink and Wood 1992; Pichon 1996; Shriar 2001). At this stage of land use, land itself has little inherent value because much of the land in the area demonstrates the qualities necessary for successful cultivation.

As land pressures mount, new and resident farmers find increasingly less suitable land in which to expand (Klepeis and Turner 2001). To compensate for the loss of the extensification option, farmers may choose to abandon a crop or switch to a new crop, or they may choose to increase the amount of labor and capital invested in an area through a process known as intensification (Boserup 1964; Netting 1993). Within swidden systems this process is typically identified as an increase in the crop cycle relative to the fallow cycle, an increase that appears to be taking place in the Calakmul area. Intensification manifests itself differently depending on economic and techno-managerial options available to farmers. Intensification develops as farmers spend more time on plots, carrying out more hand labor, such as weeding, aerating or watering plants. As their labor investment increases their output increases but with decreasing marginal returns, in some cases involving involution or stagnation (Geertz 1963; Laney 2002). Alternatively, new strategies can be adopted, usually involving added technologies. This innovative intensification (Laney 2002) typically shifts the intensification process into a new input-output state (Turner and Ali 1996), and changes attitudes toward land use. The chili story involves this kind of intensification as well, at least for the chili part of cultivation, as farmers add purchased agrochemicals and pay for mechanization.

The induced intensification thesis attempts to put all these processes together (Turner and Ali 1996) in regard to a mixed consumption-commercial production behavior among smallholders. The relative position between the two kinds of production determines the demands to which the farmer responds, either the externally driven market or internally driven subsistence needs. This response, however, is mediated by a range of factors that affect the smallholder's ability to register the demand signals: household composition, cultural morés, environmental conditions or government policy (Turner and Brush 1987; Eder 1991; Hewitt de Alcantara 1994). According to the Turner and Brush (1987) thesis, decision makers intensify production strategies when the combinations of forces creating demand make non-intensification self-destructive.

Smallholding farmers lead a dual existence, especially in frontier areas (Sanderson 1986; Netting 1993; De Janvry et al. 1997). In one respect they lead isolated lives that condemn them to poverty. In another respect, however, the farmers depend on the outside world for the subsidies they receive, the way the market operates, and how the larger national and regional societies influences their lifeways. The outside world reaps benefits from the smallholder in the form of cheap foodstuffs and raw materials.

Smallholders are different from urban dwellers due to their geographic distance from centers of power and economic production, the historical events and patterns that guarantee the secondary importance of raw material production, and in the fact that they produce food for their own consumption (Netting 1993). These differences at once weaken and embolden the peasant position. Spatial separation ensures that smallholders possess sparse knowledge of markets and policies that directly affect them (Keys 2005; Bebbington 2000). They work within the markets at a disadvantage to urban entrepreneurs and policy makers. When involved in the market, smallholders experience diminished power to negotiate terms of trade. Peasants' decisions in some sense are made separate from the day-to-day functioning of the market because of their difficulty in assessing or influencing the market. At the same time, they are not completely reliant on it, and can survive through the production of food. In the case of the Calakmul chili farmer, actual economic contact with the outside world is through the market intermediary.

METHODOLOGY

Research for this paper was carried out during 1999–2000 in conjunction with the Southern Yucatán Peninsular Region (SYPR) project (Turner II et al., 2001; Turner II et al., 2004) and involved two phases of activities. In the first phase I interviewed more than 70 local actors relevant to chili cultivation – middlemen-transporters (*coyotes*), farmers, businesses, and government and non-governmental organization agents – were interviewed to gain an overview of the history, production and marketing of chili in the region. During this process, the problem of chili marketing was identified and therefore built into the survey instrument applied to five communities of chili farmers situated along the road skirting the eastern side of the Calakmul Biosphere Reserve and crossing the hotbed of chili cultivation in the region, the so-called *zonas chileras* (Figure 10.1). The second stage of the research engaged 160 households in standardized surveys. The study selected *ejidos* by a mix of purposive sampling (in the case of the first community to cultivate chili) and by random sampling of the communities along the southern road. I selected households randomly from a complete list of *ejido* members and non-vested residents of the *ejidos*. The interviews and surveys investigated the methods, constraints and benefits of chili cultivation in addition to its regional history.

RESEARCH AREA

The southern Yucatán region, including the Muncipio de Calakmul (Figure 10.1), was once intensively occupied by the ancient Maya but was largely abandoned after the Classic Period collapse (1000 BP) until the late 19th and early 20th Centuries (Acopa and Boege 1998; Justo Sierra 1998). It is a rolling karst upland between 100–350 m.a.s.l., dominated by a monsoonal climate (Köppen Am) and seasonal tropical forest of mid-stature which loses considerable foliage during the pronounced dry season, December/January to May (Lundell 1934; Pérez-Salicrup 2004).

The modern opening of the region to farming in the late 1960s witnessed swidden or *milpa* cultivation focused on maize (*Zea mays* L.) for subsistence, although surpluses were sold to state agencies (Klepeis et al. 2002; Klepeis and Roy Chowdhury 2004). Initially older growth forest was cut, burned and cultivated for two to three years, followed by 10 to 20 years of fallow (Klepeis and Vance 2003). Once sufficient older growth was cut, successional forest became the medium of choice for annual cutting and planting (Klepeis and Turner 2001; Haenn 2005).

Land dynamics in the region changed considerably since the early 1970s. Various government-sponsored agricultural development projects collapsed along with the Mexican economy in the late 1970s (Cypher 1990; Klepeis and Roy Chowdhury 2004). Concurrently, sustained growth in the arrival of farming households arriving from elsewhere in Mexico, especially the Gulf coastal zone (Klepeis 2004), led to a regional population increase (INEGI 1996). Finally, the region was designated as a conservation zone in the 1980s and 1990s with the creation of the Calakmul Biosphere Reserve and its attendant state objective to make the Calakmul area a green exemplar (Primack et al. 1998). Various state and NGO-sponsored land practices, such as agroforestry, have been explored. The most successful option to date, however, is commercial chili cultivation, at least as measured by the amount of land devoted to it and the number of households engaged in it (Keys 2004a, 2005; Keys and Roy Chowdhury 2006).

A combination of factors encouraged farmers to enter into chili cropping. As farmers saw that those in Ricardo Payro and nearby *ejidos* were able to earn income from chili cultivation, they too began to adopt the crop. Many farmers cited multiple reasons for adopting the crop, perhaps indicating that the decision to change livelihood activities is not entered into lightly. Finally, what remains unstated in the farmer's decisions relates to their attitudes and wealth levels. Critically, the move to non-subsistence chili production constitutes a move to commercial endeavors and hence the vagaries of the market. Such a move involves a shift in economic expectations and attitudes toward risk taking, among other economic decisions (Boserup 1964; Chayanov 1965; Gudeman 1978; Hayami and Ruttan 1985; Turner and Brush 1987).

Chili first appeared in Calakmul in the community of Polo Norte (Figure 10.1). Once introduced chili cultivation spread from Polo Norte northward and southward rapidly such that by 1999 85 percent of southern Calakmul farmers cultivated chili and fully 91 percent had at least attempted cultivation (Figure 10.2). While the diffusion of new crops is common in frontier regions, the speed with which it occurred in Calakmul can be attributed to a number of factors, not the least of which is the absence of viable alternatives for income generation.

RESULTS

Farmer Types

In this paper I divide Calakmul land users into two types: subsistence farmers and commercial farmers. Commercial farmers are further discussed as two categories, swidden and mechanized commercial farmers. These distinctions are somewhat artificial; the level of market involvement of different land users more accurately resides on steps along a gradient of subsistence to commercial participation. Subsistence farmers hold an average of 44 ha of usufruct land compared to 40 ha for chili farmers.

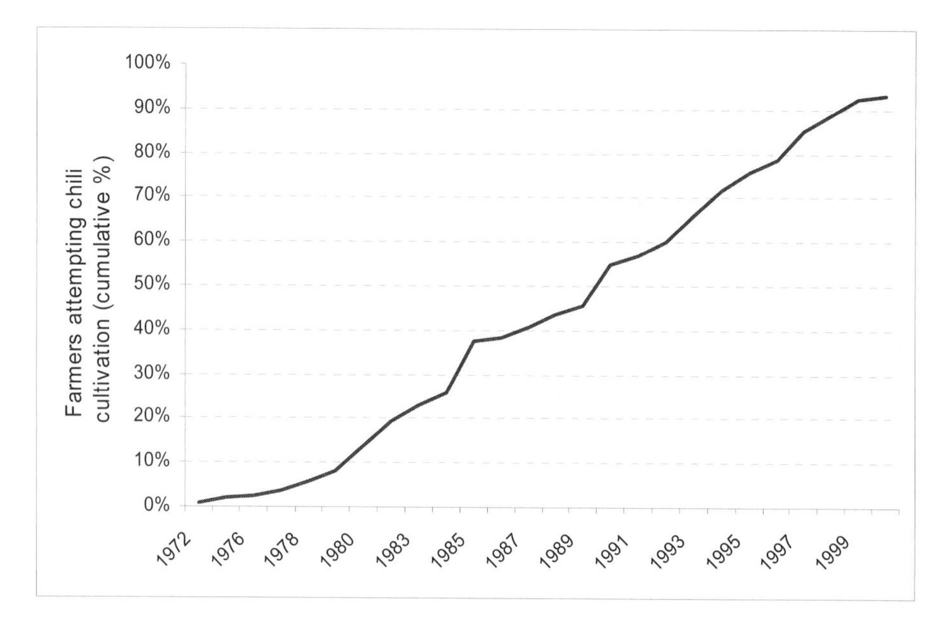

Fig. 10.2. Farmers attempting chili cultivation at least once, southern Calakmul

Twenty-five years ago, farmers in Calakmul lived by farming *milpa*. *Milpa* refers to a number of different but related technological and biological ideas. In terms of technology, *milpa* refers to a Middle American rotational, slash-and-burn agricultural system dominated by maize, beans and squash (Ewell and Merrill-Sands, 1987). To distinguish between the ideas henceforth, "*milpa*" will refer to the technological aspects of the cultivation system. "*Milpa* crops" will be used to describe the crops themselves traditionally planted in slash-and-burn – swidden – settings.

In the intervening 25 years, farmers in Calakmul shifted from the production of milpa crops for their own consumption and occasional sale to relying on chili for income while using milpa crops to feed their families. Chili today is the single most important income generator for the majority of the region's farmers, and the cash it generates supports many of the region's retail stores and supply outlets. Furthermore, chili cultivation commands a central place in governmental and non-governmental attention. In terms of the area devoted to it, chili holds second place only to milpa crops. But, in the way that it ties farmers to a wider national market and government structures, chili far outshines any other form of agriculture in Calakmul.

Agricultural Strategies

Subsistence farmers depend primarily on the cultivation of maize (*Zea mays* L.), beans (*Phaesolus vulgaris* L.) and squash (*Cucuribiticae*) to provide for household foodstuffs; this agriculture is typically termed *milpa* after both the crop and method used. Agricultural production is carried out using slash-and-burn methods that begins with plot selection in January, tree cutting if a new plot is to be cultivated, burning in May, cultivation in early June to correspond with the beginning of the rainy season and intermittent weeding and care until late September or early October when harvest occurs. Produced for household consumption, subsistence farmers occasionally sell surpluses if they exist although this occurs by accident rather than by planning.

In addition to these activities, subsistence farmers frequently engage in off-farm employment as bricklayers, field hands or in the service sector. Without this input of income subsistence farmers in Calakmul would face insurmountable hardship for household survival.

Two types of commercial farmers exist in Calakmul, those who farm commercially using slash-and-burn technology and those who farm with the use of tractors to till soil. All commercial farmers also farm *milpa* for home consumption and generally follow the same schedule for planting as that seen above. Chili cultivation follows essentially the same seasonality as *milpa* cultivation although more steps are required, especially in terms of pest management and multiple harvests.

Chili cultivation has intensified and sedentarized with respect to agroindustrial inputs. Farmers' capital needs ballooned while the area cultivated annually among chili farmers has expanded. Chili monoculture has fostered insect, plant and microbial pests (Keys 2004b). To combat these pests, increasing amounts of pesticides, fungicides, bactericides and herbicides are applied, compromising human health and ecological processes. Finally, chili cultivation exaggerates social class distinctions and cultural differentiation by demanding and encouraging a class of short-distance, day laborers while greatly increasing the wealth of a handful of smallholders.

Land-Use and Land-Cover Portfolios

One of the initial goals of this paper lay in divining whether or not commercial agriculture promised to limit land use and land cover change. This section describes the differential land cover impacts of the land uses detailed above by separating farmer type between subsistence farmers and commercial farmers. Land use and land cover are discussed in terms of subsistence farmers, swidden chili farmers and mechanized chili farmers.

In terms of land use, chili farmers cultivate more land for crops than other smallholders in the region, perhaps in an effort to expand their agricultural portfolio by increasing possible income from lower value but more securely priced crops (i.e., squash and beans). Figure 10.3 illustrates this observation. Note the increase in total area devoted to the range of crops moving from the non-chili to mechanized households. Chili farmers cultivate more land for *milpa* crops than non-chili farmers, and mechanized farmers more than non-mechanized farmers.

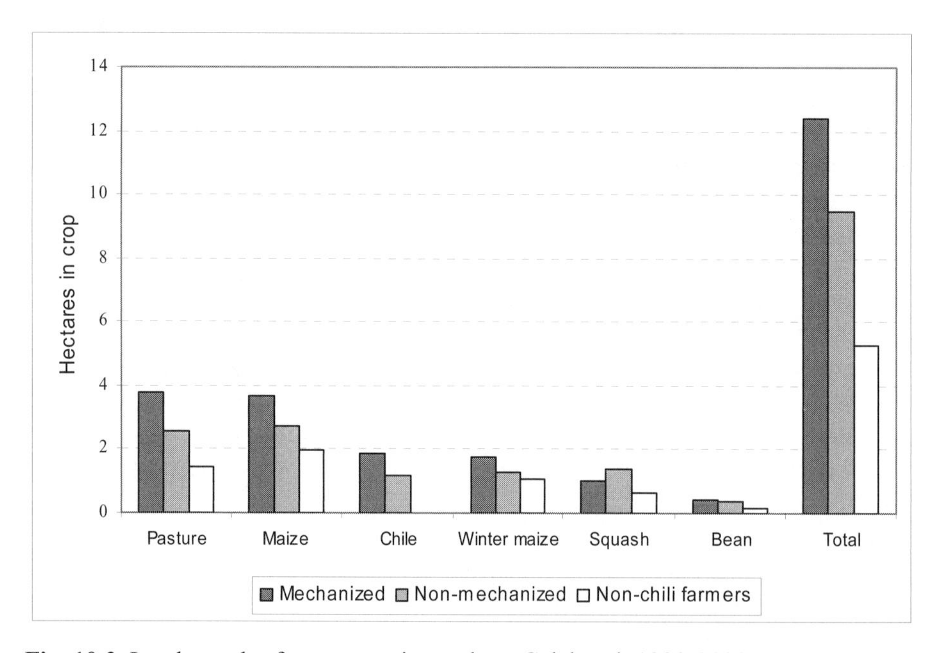

Fig. 10.3. Land-uses by farmer type in southern Calakmul, 1999-2000.

Because chili farmers cultivate more land, they might be expected to have less land in old growth forest. Figure 10.4 demonstrates the different land covers reported by farmers in the 1999 chili survey. Regrowth is any land that had been previously cultivated, but is in some state of succession. Old growth is any land that the farmer believed to have never been cultivated. Long-term agriculture describes lands that are either mechanized or pasture, and are unlikely to revert to forest in the near future. Short-term agriculture constitutes lands that are currently part of a swidden crop-fallow cycle and will be allowed to grow into forest in the near future. Mechanized chili farmers report holding the least amount of old growth land on their plots (Figure 10.4) and have the most land in uses not likely to convert back to forest in the near future, both absolutely and as a percentage of their total usufruct land. They also reported the most amount of land in regrowth and short-term agriculture.

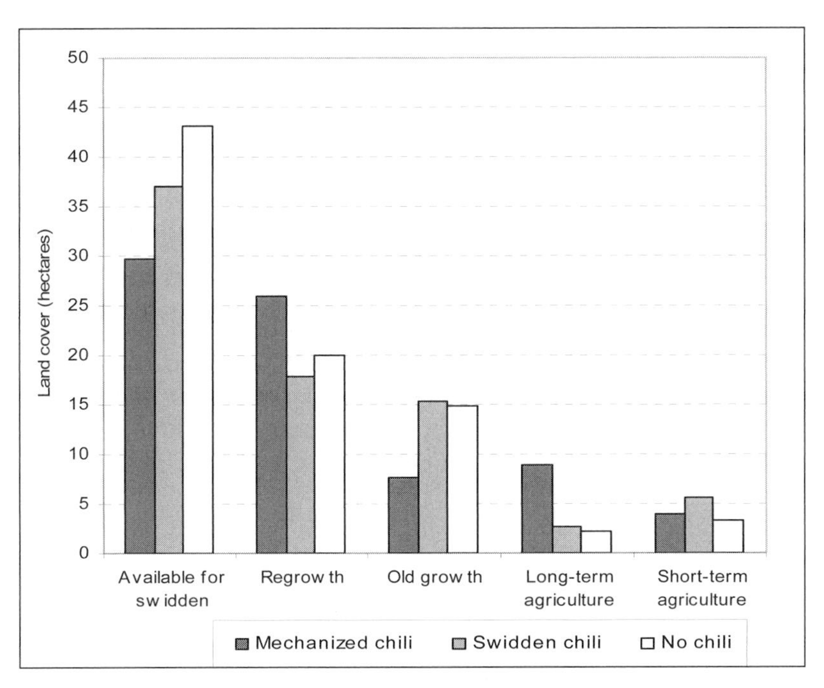

Fig. 10.4. Land covers by farmer type, southern Calakmul, 1999-2000

The reasons that chili farmers tend to use more land are varied but interrelated. Because they earn more money than most other farmers, they are more likely to hire labor from outside of their household for weeding, fumigation, clearing and harvesting, thereby expanding their agricultural portfolios. Expansion of portfolios allows farmers the opportunity to hedge against price fluctuations for their products. Increased wealth, it is assumed, affords farmers better connections to various government programs that promote access to subsidies of various kinds and to the latest experiments in cropping.

The Farm-to-Market Path: Keys to Portfolio Diversification and Agricultural Expansion

For much smallholder production in Mexico, the intermediary, or middleman, holds an especially important position (Long-Solis 1999; Keys 2005). Middlemen move product from the farm gate to national or regional markets. They are able to do so because of their possession of physical and cultural capital. The trucks, trains or other conveyances that middlemen possess physically allow them to move products. They possess sufficient capital to purchase products from farmers and sell it at an elevated price or on a consignment basis. They possess the social capital (Putnam 1993; Bebbington, 2000) that provides linkages between local leaders and the leaders of the producing companies. The middleman may provide cash to capital-poor farmers, gather product at or near the field (farm gate) and reduce the transaction cost of transportation to the farmer (Figure 10.5).

The middleman is a major translator of the worldview of the farmer to the worldview of the market (Keys 2005). Their cultural or social capital is enlarged by haggling with farmers over price or quality crops. The middleman dresses, speaks and acts in a way that the farmer understands. In some cases the middleman may speak two or more languages to better serve the producers. To be trusted, the middleman must establish reasonably honest relationships with producers and be familiar in the region. He builds up social capital in the sense that his treatment of the farmers and reputation in the region enables him to gain an advantage over other middlemen. Finally, because intermediaries travel to different farm locations and parts of the country, they witness different techno-managerial strategies for the crop in question and pass this information on to farmers elsewhere; they serve as a source of cropping knowledge.

At the other end of the farm-to-market path, the middleman must negotiate contracts, ensure delivery, arrange trucking with subcontractors, and hire and supervise lower-level middlemen. If needed, the middleman understands ways to move through the legal and commercial sectors of the market and nation. Again, at the market end of the path, he establishes trust by delivering a desired quality of product at the specified time in the needed amounts. By establishing this trust and reputation, the middleman gains an advantage with market buyers over other, competing middlemen.

Fig. 10.5. *Coyote* or middleman collecting chili harvest

The middleman is thus the link to the market, both conceptually and physically. The wholesale produce market lies somewhere between the rigid markets of stocks, bonds and commodities and the informal markets of the street. While prices depend in part on formal factors, much of the price setting for the market is based on what other sellers are selling for, the perceived demand for a product, and the relationships that the market establishes with suppliers (in this case, middlemen). If the wholesaler sets a prohibitively high price for his ware, he risks losing buyers to sellers who sell at a price even slightly below his. Likewise, the wholesaler needs to ensure continued delivery of the product. Thus, he cannot pay the middleman a price that is below the middleman's cost or does not afford him some profitability. Of course, the wholesaler must profit as well. The relationship between wholesaler and middlemen, therefore, should be perceived of as important and contingent in the chili-to-consumer supply chain.

DISCUSSION: DIFFERENT METHODS, DIFFERENT OUTCOMES

Although jalapeño is the dominant cash crop in Calakmul and the region in general, the production and sale of the peppers varies significantly from year to year. Linked to the national market through *coyotes*, chili farmers complain of inconsistent prices for their production. For example, the average price per kilogram in 1998 was N$1.43, while in 1999 it was N$2.09. These price fluctuations are linked to the state of chili production elsewhere in Mexico. States in the north and the Bajio region of Mexico have historically dominated chili production in the country and have long-standing ties to food-processing companies and intermediaries, and greater access to agricultural technologies, such as mechanization, refrigeration and chemical applications. Calakmul is best seen as a tertiary supply region, whose crop increases in value when demand is not met by the other areas, due to bad weather (e.g., killing frosts), pest outbreaks and so forth. The production in the primary regions of the country was high in 1998, and Calakmul chili prices were low. In 1999, frosts and pest attacks in the north of Mexico lowered the total national supply, and the prices were relatively high in Calakmul.

While the basic methods of chili cultivation in southern Calakmul are similar, the outcomes are radically different. Mechanized farmers appear to be deeply invested in the market and are able to profit and improve their cultivation over the years. The swidden farmers use chili as an addition to their milpa farming activities, and profit less. The precise causes and effects of these factors remain to be determined.

CONCLUSION

The arrival of migrants from Tabasco, Veracruz, and Chiapas to Calakmul set events in motion that have significantly shaped the land-use and land-cover dynamics of the southern Calakmul Municipality and will likely do so into the future. Some of these migrants landed in the region with experience in a special kind of land use not present before—commercial jalapeño production. Their arrival overlapped with the decline in large-scale projects in the region, and the dire need of the *ejido* household to find alternative ways to improve their welfare beyond that which could be gained from swidden agriculture and the sale of maize. In the absence of initial programmatic subsidies and support, a few smallholders took it on themselves to use their source community connections to *coyotes* or middlemen to draw them into chili buying in Calakmul. Thus began the chili story in the region – the final link in a chain that runs from seed to market and supplies the national market with fresh and dried peppers.

The future of chili in the region and its environmental implications are not so obvious, despite the crop's fast spread. The large flux in farm gate prices suggests that a profitable but more stable substitute might readily shift commercial practices. Beyond this, the novelty of intensive and capital input cultivation of this kind opens many questions about its ecological and economic sustainability. Chili cultivation engenders certain biophysical constraints. Pests and soil degradation appear to be growing based on the increased chemical inputs to the land. And, despite intensification, the more commercial and capital oriented the household becomes, the more of its lands are taken into cultivation and the less that remain in older growth forest. Chili cultivation, therefore, has triggered a new set of economic and ecological relationships that must be taken into account in understanding and projecting deforestation and land change in Calakmul.

The chili adoption story describes the historical process of the addition of a commercial crop into a subsistence planting portfolio and the subsequent, wide-scale adoption of the crop. All farmers in Calakmul make decisions based on both consumption and market considerations and thus intensification in the region must be understood in terms of the relative mix of types of production.

Because of the near total control on purchase that the three *coyotes* enjoy in Calakmul, the stability of prices and potential for failure challenge farmers. In the face of uncertainty the farmers respond by expanding all of their production, including in crops that in general show little market potential but that may relieve the necessity to buy foodstuffs. Because of these expanded portfolios the Calakmul Biosphere Reserve and surrounding areas experiences relatively high rates of land cover conversion, threatening the conservation efforts of the region.

REFERENCES

Achard F, Eva H, Glinni A, Mayaux P, Richards T Stibig HJ (1998) Identification of deforestation hot spot areas in the humid tropics. TREES Publications Series B. Research Report No. 4. European Commission, Luxemburg

Acopa D, Boege E (1998) The Maya Forest in Campeche, México: Experiences in Forest Management at Calakmul. In: Primack RB, Bray DB, Galletti HG, Ponciano I (eds) Timber, Tourists, and Temples: Conservation and Development in the Maya Forest of Belize, Guatemala, and México.Island Press, Washington DC 81-97

Balmford A, Green RE, Scharlemann JPW (2005) Sparing Land for Nature: Exploring the potential impact of changes in agricultural yield on the area needed for production. Global Change Biology 11(10):1594-1605

Barlett PF (1976) Labor Efficiency and the Mechanism of Agricultural Evolution. Journal of Anthropological Research 32:124-40

Bebbington AJ (2000) Reencountering Development: Livelihood Transitions and Place Transformations in the Andes. Annals of the Association of American Geographers 90(3):495-520

Boege E (1995) The Calakmul Biosphere Reserve. UNESCO (South-South Cooperation Programme) Working Paper 13, UNESCO, Paris

Boserup E (1964) The Conditions of Agricultural Growth. Aldine, Chicago

Cortina V, Macario Mendoza SP, Ogneva-Himmelberger Y (1999) Cambios en el uso del suelo y deforestación en el sur de los estados de Campeche y Quintana Roo, México. Investigaciones Geográficas, Boletín 38:41-56

Cypher JM (1990) State and capital in Mexico: development policy since 1940. Westview Press, Boulder

De Janvry A, Gordillo A, Sadoulet E (1997) Mexico's Second Agrarian Reform. Center for U.S.Mexican Studies at UCSD, San Diego

Eder JF (1991) Agricultural intensification and labor productivity in a Philippine vegetable gardening community - A longitudinal study. Human Organization 50(3):245-55.

Ewell PT, Merrill-Sands D (1987) Milpa in Yucatan: A long-fallow maize system and its alternatives in the Maya peasant economy. In: Turner B, Brush SB (eds) Comparative Farming Systems. Guilford Press, New York, pp 95-129

Galletti HA (1998) The Maya Forest of Quintana Roo: Thirteen Years of Conservation and Community Development. In: Primack RB, Bray DB, Galletti HG, Ponciano I (eds) Timber, Tourists, and Temples:Conservation and Development in the Maya Forest of Belize, Guatemala, and Mexico. Island Press, Washington DC 33-45

Geertz C (1963) Agricultural involution: the process of ecological change in Indonesia. Published for the Association of Asian Studies, University of California Press, Berkeley

Geist HJ, Lambin E (2001) What Drives Tropical Deforestation? A meat-analysis of proximate and underlying causes of deforestation based on subnational case study evidence. LUCC Report Series No. 4, CIACO, Louvain-la-Neuve

Gereffi G, Korzeniewicz M (1994) Commodity Chains and Global Capitalism. Greenwood, Westport, CT

Green RE, Cornell SJ, Scharlemann JPW, Balmford A (2005) Farming and the fate of wild nature. Science, 307(5709):550–555

Gudeman S (1978) The Demise of a Rural Economy. Routledge and Kegan Paul, Boston, MA

Haenn N (2005) Fields of Power, Forests of Discontent: Culture, Conservation, and the State in Mexico.University of Arizona Press, Tucson

Hayami Y Ruttan V (1985) Agricultural Development: An International Perspective. Johns Hopkins Press, Baltimore

Hewitt de A C (1994) Economic Restructuring and Rural Subsistence in Mexico. In: Hewitt de A C (ed) Economic Restructuring and Rural Subsistence in Mexico: Corn and the Crisis of the 1980s. Center for U.S.-Mexican Studies at UCSD, Sand Diego, pp 1-24

INEGI (1996) Conteo de población y vivienda 1995. Resumen definitiva de tabulas básicas. Aguascalientes, INEGI, México

Justo Sierra C (2001) Historia Brave de Campeche. UAC, Campeche

Keys E (2004a) Jalapeño Chili Cultivation: An Emergent Land Use in SYPR. In: Turner II BL, Geoghegan J, Foster D (eds) Integrated Land Science and the Southern Yucatan: Final Frontiers. Clarendon Press, Oxford, pp 207-220

--- (2004b) Chili Cultivation in the Southern Yucatan Region: Plant-Pest Disease as Land Degradation. Land Degradation and Development 15(4):397-409

---(2005) Market Intermediaries Link Farms to Markets: Southeastern Mexican Examples. The Geographical Review 95:24-46

Keys E, McConnell W (2005) Global Change and the Intensification of Agriculture in the Tropics. Global Environmental Change, Part A. 15:320-337

Keys E, Roy Chowdhury R (2006) Cash crops, smallholder decision making and institutional interactions in a closing frontier: Calakmul, Campeche, Mexico. Journal of Latin American Geography 5.2:75-90

Klepeis P (2004) Forest Extraction to Theme Parks: The Modern History of Land Change in the Region. In: Turner II BL, Geoghegan J, Foster DR (eds) Integrated Land-Change Science and Tropical Deforestation in the Southern Yucatán: Final Frontiers. Oxford University Press, Oxford, pp 39-60

Klepeis P, Vance C (2003) Neoliberal policy and deforestation in southeastern Mexico: an assessment of the PROCAMPO program. Economic Geography 79(3): 221-240

Klepeis P, Schneider L, Turner II BL (2002) Three Millennia in the Southern Yucatán Peninsular Region: Implications for Occupancy, Use, and Carrying Capacity. In: Gómez-Pompa A, Allen M, Fedick S, Jimenez-Osornio J (eds) The Lowland

Maya Area: Three Millennia at the Human-Wildland Interface. Haworth Press, New York

Klepeis P, Turner II BL (2001) Integrated Land History and Global Change Science: The Example of the Southern Yucatán Peninsular Region Project. Land Use Policy 18(1):272-39

Klepeis P, Roy Chowdhury R (2004) Institutions, Organizations, and Policy Affecting Land Change: Complexity Within and Beyond the Ejido. In: Turner II BL, Geoghegan J, Foster DR (eds) Integrated Land-Change Science and Tropical Deforestation in the Southern Yucatán: Final Frontiers. Oxford University Press, Oxford, pp 145-170

Laney R (2002) Disaggregating Induced Intensification for Land Change Analysis: A case study from Madagascar. Annals of the Association of American Geographers 92(4):702-726

Long-Solis J (1998) Capsicum y cultura: la historia del chilli. Mexico City: Fondo de cultura económica

Lundell CL (1934) Preliminary Sketch of the Phytogeography of the Yucatan Peninsula. In: Lundell CL, Jason RS (eds) Contribution to American Archaeology, Carnegie Institute of Washington, Washington DC, Vol.436 (12) pp 257-355

Netting RM (1993) Smallholders, householders: Farm families and the ecology of sustainable, intensive agriculture. Stanford University Press, Stanford, CA

Perez SD (2004) Forest Types and Their Implications. In: Turner II BL, Geoghegan J, Foster D. Integrated Land-Change Science and Tropical Deforestation in the Southern Yucatán: Final Frontiers. Oxford Geographical and Environmental Studies, Clarendon Press of Oxford University Press, Oxford

Pichon FJ (1996) Settler agriculture and the dynamics of resource allocation in frontier environments. Human Ecology 24(3):341-371

Primack RB, Bray D, Galletti HA, Ponciano I (1998) Timber, Tourists, and Temples: Conservation and Development in the Maya Forests of Belize, Guatemala, and Mexico. Island Press, Washington DC

Putnam R (1993) The Prosperous Community: Social Capital and Public Life. American Prospect 13:35-42

Rudel TK (2004) Tropical forests: regional pathways of destruction and regeneration in the late twentieth century. Columbia University Press, New York

Sanderson S (1986) The Transformation of Mexican Agriculture: International Structure and the Politics of Rural Change, Princeton University Press, Princeton, NJ

Schmink M, Wood CH (1992) Contested Frontiers in Amazonia. Columbia University Press, New York

Shriar AJ (2001) The dynamics of agricultural intensification and resource conservation in the buffer zone of the Maya Biosphere Reserve, Petén, Guatemala. Human Ecology 29(1):27-48

Thorner D, Kerblay B, Smith REF (1966) In: Chayanov AV (ed) On the Theory of peasant economy. Homewood, IL, Erwin RD

Turner II BL, Brush SB (1987) Comparative farming systems. The Guilford Press, New York

Turner II BL, Geoghegan J, Foster D (eds) (2004) Integrated Land Science and the Southern Yucatan: Final Frontiers. Clarendon Press/Oxford University Press, Oxford.

Turner II BL, Cortina Villar S, Foster D, Geoghegan J, Keys E, Klepeis P, Lawrence D, Macario Mendoza P, Manson S, Ogneva-Himmelberger Y, Plotkin AB, Pérez Salicrup D, Roy Chowdhury R, Savitsky B, Schneider L, Schmook B, Vance C (2001) Deforestation in the Southern Yucatán Peninsular Region: An Integrative Approach. Forest Ecology and Management 154:343-370

Turner II BL, Shajaat Ali AM (1996) Induced Intensification:Agricultural Change in Bangladesh with Implications for Malthus and Boserup. Proceedings of the National Academy of Sciences of the United States of America 93(25):14984-91.

Uc Reyes C (1999) Chile Jalapeño. In: Haggar J, Chetumal, Q. Roo (eds) Manual Agroforestal para la Peninsula de Yucatán. ICRAF, pp 4.24-4.27

Watts MJ, Goodman D (1997) Agrarian Questions:Global Appetite, Local Metabolism: nature, culture, and industry in Fin-de-siecle Agro-food Systems. In: Goodman D, Watts M (eds) Globalising Food: Agrarian Questions and Global Restructuring. Routledge, Boston, MA, pp 1-33

Whatmore S, Thorne L (1997) Nourishing Networks: Alternative Geographies of Food. In:Goodman D, Watts M (eds) Globalising Food: Agrarian Questions and Global Restructuring. Routledge, Boston, MA, pp 287-303

Wilken G (1987) Good Farmers: Traditional Agriculture and Resource Management in Mexico and Central America. University of California Press, Berkeley

CHAPTER 11 - Frameworks for Farmland Afforestation in Rural China: An Assessment of Household-based and Collective Management

Laura Ediger

Institute of Geographic Sciences and Natural Resources Research, Chinese Academy of Sciences, Beijing, China, 100101

ABSTRACT

Afforestation initiatives in China, intended to increase forest resources and improve environmental services, have been undertaken at different scales and under a variety of institutional frameworks. Both collective and household-based management systems are being utilized to encourage tree planting, with varying degrees of success. Two case studies provide examples of conversion to forest in rural southwest China, highlighting the structural flaws of mandated afforestation under the household-based agricultural production system. The use of established collective institutions to manage innovative benefit-sharing arrangements has the potential to distribute short-term costs and long-term risks while maintaining local-level control over forest resources.

INTRODUCTION

Recent afforestation projects in China have caused a substantial reduction of farming land, with both national and local government providing incentives for household investment in forest resources and the conversion of arable land to tree plantations. The goal of increasing forest area is part of a strategy to enhance the economic resources of China's forests, while providing environmental benefits through reduced soil erosion and improved infiltration of rainfall.

Total forest area has fluctuated significantly over the last several decades, with widespread destruction of forest resources continuing from the civil war period through the political and social upheavals of the early 1970s. The trend of forest depletion lessened somewhat with the shift from collective to household management of land resources in the late 1970s, when rural residents were encouraged to replant woodlots. The reform period, beginning in 1978, marked changes for both collective and state-owned forests, and forest area has increased steadily since the early 1980s (Hyde et al 2003). Stringent forest protection regulations have been instated to protect remaining forest resources, in particular under the Natural Forest Protection Plan (1998), which banned logging along much of the Yangtze and Yellow Rivers, reduced timber production of state-owned forest farms and established over 12 million ha of new plantations. Although national data on forest cover is relatively imprecise, the FAO surveys show a clear increase in forest area since 1990 (Table 11.1).

Table 11.1 China: national forest area (1000 ha).

	Area (1000 hectares)		
	1990	2000	2005
Forest	157141	177001	197290
Other wooded land	101498	97683	87615
Total	258639	274684	284905
Percent of total land area	16.85	18.98	21.15

Under current forest policy, tree-planting efforts have expanded to include household-level afforestation, with state subsidies for the conversion of both farmland and 'wasteland' (*huangshan*) to forest. The Sloping Land Conversion Program (SLCP), sometimes translated as 'Grain for Green,' was announced in 1998 and widely implemented beginning in 2001, with the objective of converting steeply sloping land to tree plantations and grassland. The goal is to reduce flooding and sedimentation in the Yangtze and Yellow Rivers, but the project has extended far beyond those watersheds.[2] The SLCP is just one of several national policies designed to increase total forest area. For example, in northern China, trees are being planted to slow the process of desertification, and as a 'green belt' to reduce dust storms. At lower levels of government as well, smaller-scale tree planting projects are being conducted for both economic and environmental objectives.

The result of these policies is that national forest area has been increasing by 1.2 percent annually, and this change extends beyond the boundaries of state forests to include existing farmland. Although much of forest policy is crafted on a national level, the process of afforestation is heterogeneous, with projects initiated at many different scales and under a diverse array of institutional frameworks. Household, collective and state management are all potential avenues for the implementation of forest expansion, and each of these has been utilized under different land use policies.

TRANSFER OF LAND USE DECISION MAKING

Decisions about land use in China have been devolved from the state to the household level over the last few decades. Historically, land use conversion was top-down in nature and involuntary. All land legally belongs to the state, with various contractual arrangements in effect to allow individuals or companies long-term use rights. This system removes the government from responsibility for management of land resources. However, ambiguities in land tenure and land use rights, combined with a history of centralized economic organization and limited avenues for dissent, allow the state the opportunity to intervene in determining land use. This is despite recent legal and institutional reforms that are designed to improve the responsiveness of government and move towards a more citizen-based approach to natural resource management.

The shift from collective to household-based resource management began in the late 1970s, with the introduction of the Household Responsibility System (*baochan daohu*). From an economic perspective, this allocation of land use rights has been largely successful, enabling a dramatic increase in grain production and increased efficiency of labor allocation (Oi 1999; Brandt et al 2002). The transfer of decision-making from collective management to indi-

[2] By 2002, 3 million ha had already been converted from cropland to forest or grassland, in 25 different provinces and autonomous regions (Bennett et al 2004).

viduals and households had rapid repercussions in rural China, improving local capacity for adaptation and the adoption of relevant technologies and crop varieties.

Afforestation projects such as the SLCP attempt to utilize the household-based framework for resource management to achieve environmental objectives. The program was intended to be entirely voluntary, relying on government subsidies to alter households' rational economic calculations in favor of tree-planting. However, this approach is tenuous because it relies on the ability of the state to successfully manipulate economic incentives for agricultural production. Cultivation of trees, either for timber or for orchard crops, requires a longer time horizon and the financial reserves to support a household in the intervening years before harvest. As the replacement of annual crops with tree crops is not immediately profitable for local residents, the environmental objectives of the national government and the economic goals of households are difficult to align. Due in part to this flaw, as well as the decentralization of tree-planting implementation, initial assessments reported that meeting quotas for conversion areas relied heavily on involuntary participation (Xu et al 2002; Xu et al 2004).

Coercive efforts by the state to encourage afforestation in effect undermine the nature of the household-based management system, in which residents make land use decisions with some assurance that they and their descendants will reap the benefits of their actions. Long-term investments in terracing, soil quality, or tree crops require confidence in the system of land-use rights. The result of state disregard for household determination of land use is that the foundations of household economic strategy are weakened, as calculations of future benefits and risks are thrown into doubt by altered estimations of the reliability of governmental institutions and the security of land use rights.

In contrast to the SLCP, which is mandated at the national level but places the responsibility for planting and management at the household level, other afforestation initiatives are designed and enforced by existing or adapted collective management mechanisms to distribute the costs and benefits of tree planting. The role of village-level collective institutions in resource management has been dramatically altered in the last few decades. Households are now allowed to allocate their own land and labor resources, and village committees typically play more of a supporting role, facilitating economic growth and organizing technical assistance and extension services. However, the institutional framework of collective governance remains, and provides an opportunity for innovative adaptations in the context of managing resources for environmental objectives.

CASE STUDIES

Western Yunnan is geographically on the margins of China, extending from Southeast Asia to the Himalayan foothills. However, it is of central importance in an environmental context because its boundaries include the upper watersheds for the Mekong, Nu and Yangtze Rivers. This characteristic has made it a target of national environmental concerns, specifically related to reducing erosion and flooding in China's central plains. The case studies described here are from two administrative villages in Baoshan County, one characterized by collective management of forest area, and the other by large amounts of farmland converted to tree plantations under the SLCP.

A combination of household surveys and informal interviews were conducted with a sample of more than 80 households, and data on land-use changes was collected using detailed historical land-use mapping exercises. Interviews were conducted between March 2004 and May 2005.

Baicai: Village-Scale Afforestation

The major afforestation event in Baicai village occurred in 1990-1991, when an area of more than 10,000 mu (over 660 ha) was converted from pasture and farmland to a pine plantation. The selected land was a large contiguous area in the hills above the village, which had previously been used for rotational planting of buckwheat and as pasture for livestock. 235 households participated in the tree-planting, on average converting around one hectare of their allocated land.

The pine plantation (known locally as 'the 10,000 mu forest') was the pet project of the long-time village party secretary, who persuaded all villagers with land in the proposed forest area to participate in converting the rotational cropland to a pine plantation. Each household contributed labor to the planting and management of seedlings, with funding for plant material and technical support provided by the forestry bureau. At the time, many residents were dubious about the prospects for a large-scale timber plantation and were reluctant to invest their land and labor.

Fig. 11.1. Baicai village with upland forest.

Baicai's upland afforestation is now managed under a unique benefit-sharing arrangement between local government and village residents (Figure 11.1). In a revival of collective resource management, the forest share-holding agreement distributed the costs of implementation, the risks of failure, and the potential for profit among the government and all of the households. The village team receives 10 percent, the administrative village 20 percent, and the farming households 70 percent of any profits. Collective mechanisms have already been used to pool money for resource protection, with the hiring of forest guards to protect against theft and fire. In late 2006, a three-year contract for thinning the pine forest was awarded, and the share system was used to allocate the rental fees. Although these initial profits from the forest are minimal, local farmers and village officials expect benefits of their investment to include both a long-term renewable income source and the improvement of local environmental services. Residents already report that soil erosion and the formation of gullies have noticeably declined.

Pingzhang: Household-Scale Afforestation

While gradual conversion of marginal land to tree plantations has occurred ever since decollectivization in the late 1970s, the most significant conversion of farmland in Pingzhang took place within the last five years under the SLCP. The SLCP program has been implemented over a wide range of land types in this village, from highly productive rice paddies to steep, rocky slopes suitable only for light grazing. National quotas are allocated hierarchically, with officials at each level (national, provincial, district, village) responsible for the distribution of their allotted conversion area. District and local forestry officials and government representatives then determine the location. In Pingzhang, forestry officials decided that participants should plant pear trees on the SLCP land. Those residents holding land use rights to the designated plots were required to participate in the planting, and held responsible for seedling survival. In one case where a farmer refused to comply, he was required to rent the selected plot to another household. For convenience, officials typically select plots that are contiguous, and readily accessed by road so that they can be easily monitored. This is one reason why productive cropland is sometimes included in the SLCP, which was meant to target marginal, highly erosive land (Figure 11.2).

Of all the households in Pingzhang, 43 percent participated in the SLCP program. This level varied among the villages, as those located at lower elevations with higher productivity and limited upland area were excluded from the program. Villages at higher elevations had much higher levels of participation (up to 77 percent of households). From the households surveyed, 55 percent were involved in the SLCP program, converting an average of 40 percent of their land to pear tree plantations.

Overall, access to cropland was significantly decreased by these programs, even as the total population remained stable (Figure 11.3). Although other economic factors and government afforestation programs have contributed to a decline in cropland, the two events mentioned above are the major factors taking farmland out of agricultural production. In both villages, households are also voluntarily converting farmland to tree plantations – both for timber and for economic crops such as walnut and eucalyptus. This type of conversion is closely linked to household characteristics such as labor availability and off-farm employment.

Fig. 11.2. Valley in Pingzhang with village in distance

INSTITUTIONAL FRAMEWORKS

The collective planting in Baicai utilizes a share-based system to distribute the costs and benefits of forest management, and capitalizes on existing governance mechanisms to implement large-scale rapid conversion to timber plantations. In this instance, the individual household has limited responsibility, beyond contributing funds for forest protection.

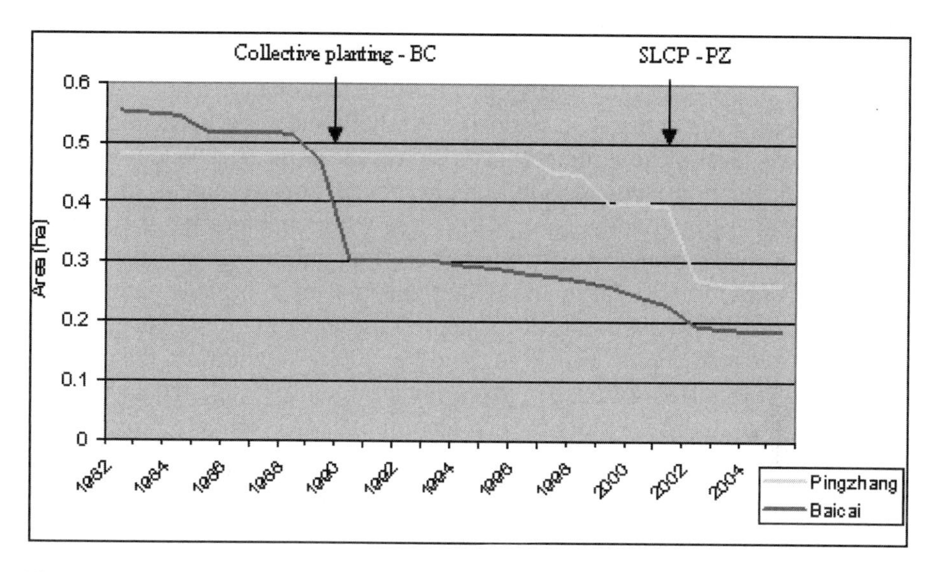

Fig.11.3. Average farmland per household, 1982-2004, based on sample surveys

By contrast in Pingzhang, while the government and forestry bureau provided the planting material and some technical assistance, the locus of responsibility is very different. The household is the focal point of management, receiving all the future benefits of timber or fruit, but also bearing the immediate costs. The role of the state is limited to initial implementation, and provision of subsidies contingent on proper management. If the seedlings die, the farmer goes without promised allotments of cash and grain. This parallels the distribution of responsibility as designed under the Household Reponsibility System (HRS) – a separation of the household and the state, with transactional support strictly legislated by defined relations.

There are two aspects in the design of household-based afforestation which present serious flaws. One is in the hierarchical system of implementation, which effectively allows local government representatives to undermine both national interests and household economic goals. The second flaw lies in the attempt to achieve national environmental goals through the household-based agricultural production system.

The first problem mentioned here for programs like the SLCP is the decentralized nature of implementation. The importance of local government in mediating between state and household goals, and the heterogeneity of government at lower administrative levels lead to wide variations in local afforestation results. In many cases, the priorities and objectives of local government officials are actually in opposition to both state and household interests. The local level of government has a strong incentive to provide evidence of successful SLCP implementation to higher levels of government, producing dramatic results of land conversion. This often works perversely, as national-level goals suffer when local officials select plots that are closer to the road (more accessible to visit), large and contiguous (easier to organize and evaluate), and relatively productive (not located on the most marginal, steep lands, and therefore less likely to fail) (Wang and Pei 2004). In addition, the production goals of the farming household are undermined, because the local government selects household plots for conversion that are relatively more accessible and productive, typically provides poor quality planting stock with minimal technical assistance, and in some cases even fails to provide the promised subsidy.

The second, more systemic problem with current state policy on household-based afforestation lies in the contradiction of state imposition of environmental protection goals onto the household economic system. With the devolution of agricultural production to the household level, the objectives of the state and the household were aligned: maximum profit (the household goal) coincided with maximum production of grain (the state goal). Various types of state interference in agriculture, in the form of grain quotas and taxes, have been gradually dismantled, and rural incomes and grain production have continued to increase.

The SLCP program is attempting to function within the HRS framework, using subsidies to provide economic incentives for changes in agricultural production. However, with afforestation, state and household goals are more complex and divergent. For the national government, the objective is primarily environmental -- to plant trees of high quantity and quality, in the most marginal and erosive areas. Conversion of land used for annual crops to tree plantations is difficult to align with household economic goals of maximizing short-term and long-term profitability, considering such factors as household life cycle stage and necessary cash flow for education and health expenses.

The SLCP program design tries to accommodate both state and household goals by providing cash and grain subsidies for the first several years after planting in order to satisfy household demand for short-term profitability long enough to reach the long-term payoffs. Effectively, the state is hoping that farmers' resourceful management will provide environmental services for the nation, as a byproduct of a diversified agro-silviculture production system. The central obstacle to success of this system may not be simply poor implementation, but a flaw with how the institutional responsibility of the program is distributed.

Under the HRS, grain production increased because decentralized resource management provided incentives for farmers to work more efficiently and productively. This immediate success was in part because of the short duration of production cycles – a typical agricultural season is only a few months long. The introduction of the HRS also caused a reduction in the large-scale, long-term investments that were made under collective management, such as terracing and the construction of irrigation systems. This is largely due to the 'time horizon' of farming households – influenced by such factors as instability of land use rights, they have focused on short-term profitability (Brandt et al 2002). This time horizon has gradually lengthened, as the last 25 years have proved relatively stable in terms of national agricultural policy, and farmers now make more investments in terrace construction and repair. However, a program like the SLCP demands a farsighted commitment, to sacrifice a portion of the household's grain harvest in exchange for the potential harvest of fruit or timber in several years' time.

Finally, another aspect of state intervention in the HRS is that required participation in state-mandated land conversion actually shortens farmers' time horizon by undermining the stability of land use rights. By demonstrating that the state is still willing to appropriate land and dictate resource management, coercive land conversion makes it less likely that farmers will voluntarily make longer-term investments.

COLLECTIVE ALTERNATIVES?

Harnessing an individualist, market-based approach for the provision of public environmental goods and services is a complex task, and there is widespread recognition that individuals have little incentive to forgo short-term economic gain for long-term collective benefits. A number of mechanisms and institutional frameworks have been devised in different contexts to overcome this tendency and introduce collective oversight of individual resource management behavior (Ostrom 1990).

Village-level elections in China have begun to alter the relationship between residents and officials, with more accountability at the lowest levels of governments and increased expectations from local constituents (O'Brien and Li 2000). While this increase in local accountability of village officials is an important influence on local land use, it is the combination with older, collective systems of management in the case of longer-term tree crops that is particularly intriguing. This represents an integration of reform-era governance practices with historical collectivist institutions that have been largely abandoned by the state in favor of individual or household-based management regimes.

The suitability of this type of management depends in part on the resource to be managed. One example of collective management of wild mushrooms in a nearby village was corrupt and badly managed, and local residents demanded that land use rights be returned to the original households (Dong 2006). However, for long-term investments such as timber plantations, where the short-term economic costs far outweigh the immediate benefits, collective institutions can play a crucial role in distributing the current costs and eventual profits. As the environmental objectives underlying national afforestation initiatives rely heavily on long-term investments, the institutional framework is a key determinant of success or failure. Returning crop management to households and an individualist approach to agricultural production produced one of the most impressive increases in grain production in the last few decades, but this framework is structurally unsuited to fulfill similar expectations for afforestation.

The case of Baicai's collective forest is neither universally applicable nor without its own systemic flaws. Local land use management systems vary widely, with one study identifying nearly 30 different forest management schemes among 40 villages in Yunnan and Fujian provinces (Qiao 1997, cited in Brandt et al 2002). Although policy statements by the State Council may give the impression of rural homogeneity and centralized prescription of rural economic policy, the variation among townships and even villages within the same township is striking. The success of a particular afforestation scheme often depends on a charismatic and trusted local leader (as in Baicai), or in preferential assistance from district forestry bureaus. A beneficial outcome for both households and local governments is not guaranteed: in Baicai there are concerns about the quality of timber in the collective plantation and the technical capacity for forest management. However, the existence of recognized institutions for collective management is a unique opportunity, and the case described here shows that these institutions have the potential to be adapted as an organizational system for innovative projects.

CONCLUSION

Recent trends of afforestation in China may strike observers as a model of forest protection and a beneficial use of centralized power to coordinate the improvement of public environmental services. The statistics of forest area expansion and the sheer scale of tree-planting are certainly impressive. In reality, afforestation initiatives over the last two decades have been diverse and occur in the context of a range of institutional frameworks, from collective management of timber plantations to household subsidies for conversion of small plots.

State-mandated land conversion in the context of household management is flawed not only due to the misalignment of state and household-level objectives, but due to the unique institutional context of resource management in China. A system that is highly centralized in terms of decision-making and planning but decentralized in terms of implementation faces significant obstacles in conducting large-scale change at the local level. This is due in part to the remaining emphasis on top-down accountability, and the reliance on quantitative results for project evaluation. In addition, it is a structural result of local implementation without local accountability, where lower-level officials are meeting mandated outcomes from above, with no stake in the local impacts of a project. Finally, high levels of corruption ensure that well-funded state projects attract special attention from officials at all levels, as an opportunity to supplement low salaries.

For the long-term investment that is required for reducing erosion and improving hydrological stability, a program that insufficiently coincides with farmers' existing incentives will be ineffective. A program like the SLCP ironically reduces farmer incentives for long-term investment by undermining the growing stability of land tenure in China, making it more difficult for farming households to make long-term production plans. Instead of focusing environmental protection efforts on the household scale, it may be more effective for government policy to take advantage of local collective institutions, which can realign farmers' incentives by distributing the costs and risks of the investments over a larger group.

REFERENCES

Bennett J, Zhang L, Dai G, Xie Chen, Zhao J, Liang D, Liang Y, Wang Xuehong (2004) A Review of the Programme for Conversion of Cropland to Forest and Grassland. Research Report No. 2, Sustainable Land Use Change in the Northwest Provinces of China Research Reports, Australian Centre for International Agricultural Research. ISSN: 1449-7433

Brandt L, Jikun H, Guo L, Scott R, (2002) Land rights in rural China. Facts, fictions and issues, The China Journal 47:67-97

Dong M (2006) Initial report of commodity chain analysis of selected NTFPs in Baoshan Prefecture. Research report, World Agroforestry Centre

Food and Agriculture Organization (FAO) (2005) Global forest resources assessment 2005 China. Country Report 051, Rome, http://www.fao.org/forestry/foris/webview/common/media.jsp?mediaId=8859&geoId=102

Hyde WF, Brian B, Jintao X (2003) China's Forests: Global Lessons from Market Reforms. Resources for the Future Press, Washington DC

O'Brien KJ, Lianjiang L (2000) Accommodating democracy in a one-party state: introducing village elections in China. The China Quarterly 162:465-489

Oi J (1999) Two decades of rural reform in China: An overview and assessment. The China Quarterly 159:616-628.

Ostrom E (1990) Governing the commons: the evolution of institutions for collective action. Cambridge University Press, Cambridge

Qiao F (1997) Property rights and forest land use in southern China. M. Sc. thesis, Graduate School of Chinese Academy of Agricultural Sciences, Beijing

Wang Q, Pei X (2004) Study on ecological impact of environmental policy in China's western development : a case study of converting the land for forestry and pasture. Proceedings of the 12th International Conference on Geoinformatics, University of Gävle, Sweden, http://www.millenniumassessment.org/documents/bridging/papers/qiao.wang.pdf

Xu J, Eugenia K, Thomas AW (2002) Implementing the Natural Forest Protection Program and the Sloping Land Conversion Program: Lessons and Policy Recommendations. China Council for International Cooperation on Environment and Development

Xu Z, Michael T, Bennett RT, Xu J (2004) China's Sloping Land Conversion Program Four Years on: Current situation, pending issues. International Forestry Review 6(3/4):317-326

CHAPTER 12 - Cost-benefit Analysis of Two Models of Agroforestry Systems in Rondônia, Brazil

Keila Aires

Nicholas School of the Environment and Earth Sciences, Duke University, Durham, NC, USA 27705

ABSTRACT

The Amazon Forest is a known wealth of fauna, flora, fresh water and many other ecosystem services. However, the Forest has been hardly damaged by deforestation caused by other land uses such as pastureland, logging and agriculture. The Forest is present in nine Brazilian States in a region occupied by more than 20 million people, one of the poorest regions in the country. Hence, in order to decrease land uses that cause deforestation it is necessary to promote those that foster forest conservation and enhance the local economy, such as agroforestry. This study analyzed the financial viability of two agroforestry systems. It was found that the studied agroforestry systems are profitable and can be recommended as profitable and sustainable land use. The results include high NPV, IRR and low payback for both systems. The costs of implementation and maintenance proved not to be compromising, even when a sensitivity analysis was applied to verify the results in case of 20 percent lower productivity.

INTRODUCTION

The Amazon forest contains 20 percent of the animal and plant species of the planet (Wilson 1999) and it is a large source of available fresh water. It is also home to more than 20 million people (IBGE 2000). Deforestation is expanding rapidly, causing loss of biodiversity, soil erosion, water pollution, silting of rivers, and influencing on climatic change.

The Brazilian Amazon is divided into nine states (Acre, Amapá, Amazonas, Mato Grosso, Maranhão, Pará, Rondônia, Roraima and Tocantins). The State of Rondonia has been one of the most affected by deforestation resulting from three principal activities: logging, agriculture and cattle grazing. (Pfaff 1999; Pedlowski et al. 1997; Kaimowitz and Angelson 2004; Rudel 2005; Nepstad et al. 2006).

The implementation of agroforestry systems can assist in reducing land clearing, and is considered an important option for sustainable development in the Amazon (Dubois 1979; Budowski 1984; Current et al. 1995; Nair 1993; Medrado et al. 1994). The objective of the work reported in the chapter was to verify the cost benefit of two models of agroforestry systems adopted in the Rondônia. Demonstrating financial viability of agroforestry systems weakens the widespread argument and belief that slash-and-burn and pasture is the only profitable land use in the Amazon. Showing profitability of agroforestry can encourage producers to shift their current farming techniques to the use these systems.

LAND USE IN THE STATE OF RONDÔNIA

In order to understand the socio-economic environmental problems present in the State of Rondônia, it is necessary to understand the context in which the State developed. The accelerated expansion of the Rondônia frontier between 1960 and 1980 was accompanied by diverse socio-environmental problems, including conflicts over land and other natural resources, high rates of deforestation, soil deterioration, as well as disorganized urban growth (Godfrey and Browder, 1997; Fearnside 1990). However, the most important environmental impact of the rapid frontier expansion in Rondônia was deforestation, especially for cattle ranching (Fearnside and Ferreira 1984; Walker and Homma 1996; Arima and Uhl 1997). According to the Instituto Nacional de Pesquisas Espaciais[1] (INPE), in 1978 the area deforestated in Rondônia was estimated at 121,700 hectares or 1.8 percent of the State's total area. By 1980, the deforested areas comprised 3.1 percent of the State, and by 2000 24.5 percent.

This process, resulting from agricultural and cattle ranching practices, began with a system of migratory agriculture. Migratory agriculture consists of traditional subsistence agriculture where areas of forest are cut, burned and sowed with subsistence crops, such as rice, beans and corn, for a maximum period of two years and afterwards abandoned, or turned over to pasture (Caviglia and

[1] The Brazilian National Institute for Space Research.

Kahn 2001). Temporary cultivation was subsequently transformed into permanent cropland or pasture (Carvalho 1995).

With the conversion of land formerly under migratory agriculture into pasture, Rondônia soon developed a high concentration of cattle ranching properties, an activity that now plays an important role in the state's economy. According to the Agriculture and Livestock Census of 1996, there were more than 6 million head of cattle in Rondônia, with 33 percent of the total area of the State occupied by pasture (Fearnside 1996).

Cattle raising in the State of Rondônia has rapidly developed for various reasons (Arima and Uhl 1997; Egler 2001; Andersen et al. 2002): 1) livestock is considered a liquid investment which can be sold in times of crisis or opportunity, and the sale can be deferred without great financial losses; 2) livestock requires little labor, thereby reducing maintenance costs; 3) the cost of creating pasture after opening the land for cultivation is low compared to other crops; 4) livestock provide other products that add to producers' revenues, such as milk, hides and calves; 5) livestock is considered a safe investment for middle class families; and 6) production of livestock is not dependent on the availability of roads, because herds can be moved in droves. After several years, however, particularly in areas of poor soils that are subject to erosion, pastures become unproductive, resulting in soil deterioration, silting of watercourses and consequently there is further destruction of primary forests as farmers seek for more land for pasture (Fearnside 1990; Nepstad et al. 1996; Fujisaka and White 1998; Chomitz and Thomas 2001).

The objective of promoting the use of agroforestry systems is not only to create incentives for preservation, but also to offer an alternative that is profitable when compared to migratory agricultural production and extensive cattle ranching. To improve the way in which people manage, maintain and even rehabilitate degraded areas may appear to be a magic formula that is impossible to apply, yet there are characteristics applicable to agroforestry systems that can maintain the productive capacity of the land and enhance it even to the extent of rehabilitating areas that have suffered the effects of slash-and-burn practices (Dubois et al. 1996).

AGROFORESTRY SYSTEMS STUDIED

Study Area

The study area comprises 100 hectares and was planted in 1997. At the time of the study, the plantations had been in operation for six years. Half of the study area was occupied by a consortium cultivating cacao (*Theobroma cacao* L.), coffee (*Coffea canephora* Pierre ex. Froenher) and teak (*Tectoma grandis* L.F.) and the other half for a consortium growing cacao, peach palm (B*ractis gasipaes* Kunth.) and *freijo* (Spanish elm, *Cordia alliodora* (Ruiz & Paz.) Oken). Both are located in the municipality of Ouro Preto D´Oeste in the State of Rondônia, within the Executive Planning Commission for Cultivation of Cacau (CEPLAC) Experimental Station.

The productivity coefficients, planting and other costs, and market values of the yields were provided by the following state institutions: Comissão Executiva do Plano para Lavoura Cacaueira[2] (CEPLAC); Associação de Assistência Técnica e Extensão Rural[3] (EMATER/RO); and Empresa Brasileira de Pesquisa Agropecuária[4] (EMBRAPA). The commercial values of timber were calculated using the average of the annual prices of timber (teak and freijo) provided by the chamber of commerce in Porto Velho, Rondonia (price of timber for export)..

Given the distribution of the components over the time of cultivation, the agroforestry consortia models used for this study are characterized as a spatially-mixed. They are also simultaneous systems, meaning that the crops were planted simultaneously in the same area. The World Agroforestry Center (ICRAF) classifies this type of forestry consortium as agri-silvicultural as it combines trees or bushes with cultivated crops. CEPLAC-Rondônia classifies them as interspersed permanent mixed systems.

Cacao, coffee and teak agroforestry system

According to Almeida (2002), this agroforestry system is predominant in the Rondônia, uniting the two principal perennial crops cultivated in the state for

[2] Executive Planning Commission for Cultivation of Cacau
[3] Association for Technical Assistance and Rural Expansion
[4] Brazilian Association for Livestock Research

agro-economic strategic reasons. This agroforestry consortium model also combines two commodities that might be susceptible to price fluctuations, which guarantees the consortium's economic sufficiency in periods of unfavorable market conditions for one of the two crops.

The area of study consists of three strips of cultivation, of which two are cacao, interspersed with one of coffee. In each strip of cacao, ten rows are planted in a spacing of 3 x 2 m, while the coffee cultivation consists of eleven rows with a spacing of 3 x 1 m. Between these planting zones there is a row of teak trees, with a spacing of 2.5 m between plants and 3 m between the adjacent strips. Teak is used to provide permanent lateral shade for the cacao plants, which are sensitive to sunlight. The rows of teak are spaced 33 m apart when intercropped with cacao, or 36 m apart when intercropped with coffee. Banana trees (*Musa sp.*) are planted with a spacing of 3 x 4 m. The banana trees have an important function in both consortia studied: they furnish shade for the cacao plants in the first three years of growth, and they are an important source of revenue in the period following initial investment.

Teak, the tree component of the system, is a southeast Asian species that is characterized by the ease of cultivation, timber durability, a rectilinear trunk, fast growth in the region and fire resistance. For these reasons, the teak wood has a high market value, e.g., it is used in the production of fine furniture, quality set squares, interior design and marine construction.

The combination of the soil and climatic conditions in Ouro Preto D'Oeste produces teak greater than 8 m height and with a diameter at breast height (dbh) of approximately 11.6 cm after the first three years of cultivation, and 12.5 m and 17.0 cm, respectively, after five years. It is estimated that trees will reach a commercial size 22 years after planting, when they should have a dbh >40 cm and a trunk height exceeding 15 m (Alameida 2002). The tree densities of this system are 975 cacao plants/ha, 1062 coffee plants/ha and 117 teak trees/ha.

Cacao, peach palm and *freijo* agroforestry system

This consortium consisted of planting strips comprised of ten rows of cacao plants, each of 3 x 2.5 m, alternated with three rows of peach palms, with a spacing of 2 x 1.5 m in a north-south direction. Between the intercropped species there is a distance of two meters. The triple rows of peach palms are placed 31 m apart. Peach palms are primarily grown for the cultivation of palmito (peach palm heart). This palm is spinelesss, which facilitates plant handling. In addition, this planting structure provides the cacao plants with shelter from the wind. Temporary shade for the cacao plants is provided by banana trees planted planted on rows of 3 x 2.5 m (Almeida 2002).

The first harvest of the peach palms should take place 24 months after planting, with further harvests every six months. Permanent shade for the cacao is provided by *freijo* which is planted in the cultivation zone of the cacao plants, with four trees in an area of 12 x 10 m. A distance of 3.5 m is maintained from the first row of peach palms. *Freijo* is characterized by a low crown density and rapid initial growth. Its wood is used for a variety of purposes, and has a good price in both external and internal markets. The planting densities for this system are 1145 cacao plants/ha, 586 peach palms/ha and 88 *freijo* trees/ha.

During the first four years after planting, agroforestry systems set up on soils in the initial stages of deterioration in the Estação Experimental da CEPLAC (CEPLAC Experimental Station) in Ouro Preto O'Este, had an average growth rate of 188 and 101 mm/month in height, and 9 and 12 mm/month in trunk circumference, for *freijo* and peach palm respectively (Almeida, 2002).

Estimated productivity

The productivity levels from the sixth year of production onwards are based on data obtained from CEPLAC/RO taken from studies in different regions of Amazonia, together with field observations in rural areas in the interior of Rondônia.

It can be observed that the productivity of coffee in the twelfth and thirteenth years is zero (Table 12.1). This is due to twelfth-year pruning of coffee plants, which results in inactivity in the thirteenth year. In the fourteenth year production is considered to be equal to that of the first year, while in the fifteenth year production is equal to that of the second year, and in by the sixteenth year the production is equal to that of the third year when it again stabilizes.

Teak production was projected to be assessed only in the final cut, as it is unnecessary to thin teak because the spacing adopted between trees allows for its growth until the end of the productive period. A mortality rate of 5 percent is accepted for the teak trees, and the final production, as shown in Table 12.1, is evaluated in terms of logs with inside bark.

The production of *freijo* is evaluated over three phases: 1) to the first commercial thinning, i.e. removal of trees that are not following the anticipated growth pattern, in the tenth year; 2) the second commercial thinning in the fifteenth year; and 3) the final cut in the twenty-fifth year. Its production is also evaluated in terms of logs with inside bark, as shown in Table 12.1.

Table 12.1 – Estimated production per consortium (Value US$ per 50 hectares)

Consortium Cacao, Coffee & Teak					**Consortium Cacao, Palm & Freijo**			
Year	Cacao	Coffee	Banana	Teak	Cacao	Palm Heart	Banana	Freijo
1			39,000				39,000	
2			58,500			30,000	45,000	
3	9,000	12,500	115,000		10,500	45,000	115,000	
4	18,000	18,500			42,000	115,000		
5	36,000	25,000			63,000			
6	52,500	25,000			63,000			
7	52,500	25,000			63,000			
8	52,500	25,000			63,000			
9	52,500	25,000			63,000			
10	52,500	25,000			63,000			289
11	52,500	25,000			63,000			
12	52,500				63,000			
13	52,500				63,000			
14	52,500	12,500			63,000			
15	52,500	18,500			63,000			692
16	52,500	25,000			63,000			
17	52,500	25,000			63,000			
18	52,500	25,000			63,000			
19	52,500	25,000			63,000			
20	52,500	25,000			63,000			
21	52,500	25,000			63,000			
22	52,500	25,000			63,000			
23	52,500	25,000			63,000			
24	52,500	25,000			63,000			
25	52,500	25,000		8,350				1873

Volumes: Cacao (dried/kg/ha); Coffee (kg/ha); Bananas (kg/ha); Palm Hearts (unit/ha); Freijo (m^3/ha)

COST AND BENEFITS OF THE AGROFORESTRY SYSTEMS STUDIED

Initial Investments

The costs include the initial investments made in the construction of buildings for drying cacao and coffee, as well as costs of plating and maintaining the two agroforestry systems during the two first years.

This study does not take into consideration the purchase value of the land, as the farmers already owned the property at when they formed the consortia. It should be noted, however, that the exclusion of the land acquisition costs directly influences the calculation of the internal rate of return (IRR) and the net present value (NPV) which are discussed below. These would change, for instance, if the land values were to have increased after purchase. In addition, the construction of administrative buildings and fences was not taken into consideration. The decision to ignore these values is justified by the fact that the agroforestry consortia are, generally, established by farmers who already own rural properties but wish to change the production system, or that have land that does not possess the requisite conditions for growing rice, beans or corn.

The total value of the initial investments in the cacao-coffee-teak consortium was US$ 82,800 and the total value of the initial investments in the cacao-peach palm-teak consortium was US$ 83,960[5].

Total costs

The annual costs of the consortia included soil preparation, seedlings acquisition, planting, cultural treatments, harvest, supplies, maintenance and labor. The total for each consortium per annum is shown in Table 12.2.

Estimated revenue:

The projection for the production volume in the cacao-coffee-teak consortium was made on the basis of a density of 975 cacao plants/ha. The production volume was estimated at 180 kg/ha of dry cacao for the first year of pro-

[5] The conversion rates applied was R$ 2.87 to US$1.00

duction, 360 kg/ha in the second year, 720 kg/ha in the third year and 1,050 kg/ha in the fourth through to the last year of production. The estimated price for the sale of cacao was US$ 1.07/kg, based on the local market price in May 2005The cultivation of coffee was projected on a basis of a density of 1,062 plants/ha and an estimated volume of production of 250 kg/ha in the first year of production, 376 kg/ha in the second year, and 500 kg/ha from the third throughout twelfth year, when the plants are pruned, productive cycle reinitiated and maintained until the twenty-fifth year. The estimated price of coffee was considered to be US$ 0.58/kg.

The projection for the production of bananas was based on a density of 975 plants/ha. Production was established at 7,800 kg/ha in the first year, 15,000 kg/ha in the second year and 23,000 kg/ha in the third year, when the crop was eliminated from the consortium. The principal objective of the banana trees was to provide temporary shade for the cacao plants, so production was evaluated only up to the consortium's third year

The teak component has a density of 117 plants/ha and a productive cycle of twenty five years. Given the spacing used by the consortium for this component, thinning is unnecessary. Consequently all production occurs at the end of the twenty-five year cycle, when it is estimated that each tree will yield 1.5 m^3 of timber. A mortality rate of 5 percent was applied. The price applied to the projection was 500 US$ per m^3 of logs with bark.

For the cacao-peach palm-*freijo* consortium, the cacao component was estimated to have a density of 1145 plants/ha, with a projected production of 210 kg/ha of dry cacao in the first year, 420 kg/ha in the second year, 840 kg/ha in the third year, and 1240 kg/ha from the fourth year to the end of the 25-year cycle.

The estimated production of the peach palm was based on a density of 586 plants/ha, allowing for a projected production of 600 hearts of palm/ha in the first year of production, 840 hearts/ha in the second year and 1240 hearts /ha from the third year to the end of the 25-year cycle. The price applied in the projection was US$ 0.32 per heart of palm.

The banana component of this consortium was based on a density of 975 plants/ha; 7800 kg/ha was projected for the first year of production, 15,000 kg/ha for the second year, and 23,000 for the third year after which the crop is eliminated from the consortium.

Consortium Cacao, Coffee & Teak				Consortium Cacao, Palm & Freijo		
Year	Coffee	Cacao, Teak Banana	Total Costs	Peach palm	Cacao, Freijo & Banana	Total Costs
1	203.6	408.36	30598.26	430.91	523.12	47,701.48
2	168.87	587.94	37840.59	70.67	293.94	18,230.23
3	186.29	665.92	42610.63	74.05	350.91	21,248.00
4	193.26	664.18	42,871.95	74.05	350.91	22,293.29
5	196.74	667.67	43,220.38	74.05	350.91	22,467.51
6	196.74	695.54	44,614.11	74.05	350.91	23,861.24
7	196.74	695.54	44,614.11	74.05	350.91	23,861.24
8	196.74	695.54	44,614.11	74.05	350.91	23,861.24
9	196.74	695.54	43,917.25	74.05	350.91	23,861.24
10	196.74	695.54	43,917.25	74.05	350.91	23,861.24
11	196.74	695.54	43,917.25	74.05	350.91	23,861.24
12	369.48	695.54	52,554.01	74.05	350.91	23,861.24
13	494.65	695.54	42,000.87	74.05	350.91	23,861.24
14	179.32	695.54	42,349.30	74.05	350.91	23,861.24
15	189.77	695.54	42,871.95	74.05	350.91	23,861.24
16	193.26	695.54	42,871.95	74.05	350.91	23,861.24
17	196.74	695.54	43,220.38	74.05	350.91	23,861.24
18	196.74	695.54	43,220.38	74.05	350.91	23,861.24
19	196.74	695.54	43,220.38	74.05	350.91	23,861.24
20	196.74	695.54	43,220.38	74.05	350.91	23,861.24
21	196.74	695.54	43,220.38	74.05	350.91	23,861.24
22	196.74	695.54	43,220.38	74.05	350.91	23,861.24
23	196.74	695.54	43,220.38	74.05	350.91	23,861.24
24	196.74	695.54	43,220.38	74.05	350.91	23,861.24
25	196.74	3456.35	182,654.53	74.05	1,008.94	54,149.74
Total Costs	5326	19665.68	1213801.54	2204.73	9546.02	639,453.81

The tree component *freijo* was estimated to have a density of 88 plants/ha and provides a productive cycle of 25 years, with two commercial thinnings projected in the tenth and fifteenth years, with final harvest taking place at the end of the 25-year cycle. The mortality rate was estimated at 10 percent. A cut of 30 percent was projected for the first year and of 20 percent for the second. The final harvest of the remaining was projected at end of the 25-year cycle .

The value used for establishing projected revenue from *freijo* was US$ 104/m^3. Table 12.3 shows the annual revenue for each of the consortia.

Table 12.3 – Expected annual revenue for each consortium:

Year	Consortium Cacao, Coffee & Teak	Consortium Cacao, Peach, Palm & Freijo
1	67,944.25	67,944.25
2	130,662.02	130,662.02
3	217,280.49	220,919.86
4	30,256.45	35,560.98
5	53,181.18	64,609.76
6	70,888.50	86,073.17
7	70,888.50	86,073.17
8	70,888.50	86,073.17
9	70,888.50	86,073.17
10	70,888.50	86,073.17
11	70,888.50	86,073.17
12	70,888.50	86,073.17
13	56,341.46	86,073.17
14	56,341.46	86,073.17
15	63,614.98	158,482.93
16	67,280.84	86,073.17
17	70,888.50	86,073.17
18	70,888.50	86,073.17
19	70,888.50	86,073.17
20	70,888.50	86,073.17
21	70,888.50	86,073.17
22	70,888.50	86,073.17
23	70,888.50	86,073.17
24	70,888.50	86,073.17
25	4,239,388.50	281,909.41
Total Revenue	6045619.13	2509406.27

Internal Rate of Return (IRR)

Using the costs and the revenues mentioned above, an IRR of 90.15 percent was established for the cacao-coffee-teak consortium. The calculation of the IRR took the following variables into account: total revenue, profits, opportunity cost, initial investment and life span of the project. The IRR expressed as a percentage demonstrates the advantages in pursuing this activity when one observes the margin offered in relation to opportunity cost; in this case the rate of comparison being an indicator of the Brazilian Central Bank known as the SELIC rate, and equal to 18 percent in May 2003.

An IRR of 93.4 percent was calculated for the cacao-peach palm-*freijo* consortium. This value reflects the same variables as above: total revenues, profits, opportunity cost, initial investment and the life span of the project. The results display a slight advantage for the cacao-peach palm-*freijo* consortium over that of the cacao-coffee-teak project, although both demonstrated significant commercial viability.

Sensitivity analysis:

The sensitivity of investment was analyzed, given the possibility of 10 and 20 percent reductions in revenues. In these scenarios, the cacao-coffee-teak project showed significant liquidity; for a projected 10 percent reduction in revenue, the IRR of this consortium dropped to 75.96 percent and for a 20 percent reduction the projected IRR was 60.59 percent. Both are still financially highly attractive.

The same possibilities of a reduction of 10 or 20 percent in revenues were applied to the cacao-palm heart-*freijo* consortium. In this case, the IRRs are 81.74 and 69.79 percent respectively. These are slightly higher than the indices of the cacao-coffee-teak consortium and they also highly attractive

Net Present Value (NPV)

An analysis of the NPV for the cacao-coffee-teak consortium, with a discount rate of 18 percent for opportunity cost, indicates a profit of US$ 253,450 – a profit rate of 355 percent. For the cacao-peach palm-*freijo* consortium, the NPV profit discounted for opportunity costs was US$ 333,600, representing a profit rate of 452 percent. In considering these results it is important to highlight that the consortia presented higher costs than normal monocultures would. However, they also presented higher profits per hectare than monocultures.

Profitability and payback

The operational profitability and investment recuperation (payback) period for the cacao-coffee-teak consortium suggest a recovery of investment within sixteen months, taking into consideration that the production cycle of banana trees cultivated in the beginning of the consortium (up to the third year) over and above recovering investment costs generates resources for the development of the project. The operational profit of the cacao-peach palm-*freijo* consortium suggests an investment recovery within seventeen months. A summary of the profitability for both consortia is presented in Table 12.4.

Table 12.4 Profitability summary

	Consortium 01	Consortium 02
Indicator	Cacao -Coffee -Teak	Cacao -Peach Palm -Freijo
Initial Investments	82,800.00	83,960.00
Total Costs (US$)	1,213,975.78	650,521.44
Internal Rate of Return (%)	90. 15%	93. 4%
Net Present Value (US$)	253,450.00	333,600.00
Payback	16 months	17 months

CONCLUSIONS

In this study I analyzed the costs and benefits of two models of agroforestry system cultivated in the State of Rondonia, Brazil. The study aimed at verifying the financial viability of these agroforestry systems and it found that they are viable and highly profitable. The strong financial indicators found in this study show that agroforestry system is a farming technique that beyond been very environment friendly it is also lucrative. These findings enforce the idea that agroforestry systems constitute a good option for local farmers who currently use either slash-and-burn farming or cultivate pasture as monoculture.

Influencing Amazonian farmers to practice a more sustainable agriculture has consequences beyond local economic gains. It influences the region's land use pattern that has been for decades driving deforestation in the Amazon Forest. By using agroforestry systems farmers would be able to make better use of already deforested lands and, consequently decreasing the pressure on pristine forests.

Nevertheless, it is important to highlight that the success of the agroforestry systems use in the Amazonian Region also depend on factors as important as profitability, such as availability of technical and financial support, market, effective distribution modals, and land tenure.

The results found that the cacao-peach palm-freijo agroforestry system presented slightly better financial indicators than the cacao-coffee-teak system. Although the cacao-coffee-teak system had a higher cash flow, it also had higher costs which decreased its profitability. The lower profitability of the cacao-coffee-teak system was also influenced by the revenue distribution, as it had the income from teak timber only at the end of the 25th year, while the cacao-peach palm-freijo agroforestry system had its revenue distributed along the total time of the project, due to the thinnings applied to the freijo trees.

The slight difference between the two agroforestry system profitability show that it is important to keep the flow of income throughout the life time of the project. The constant flow of income also influences the decision of some small scale farmers, since they rely on their lands as the only source of income. Consequently, for further studies it is recommended the inclusion of more seasonal crops and the commercialization of other non-timber products coming from the tree components, such as seeds and carbon sequestration.

REFERENCES

Andersen LE, Granger CWJ, Reis EJ, Weinhold D, Wunder S (2002) The Dynamics of deforestation and Economic Development in the Brazilian Amazon. Cambridge University Press, Cambridge, UK

Almeida MCVC (2002) Sistemas Agroflorestais com o cacaueiro como alternativa sustentável ao desmatamento no Estado de Rondônia. IV Brazilian Congress of Agroforestry

Arima EY Uhl C (1997) Ranching in the Brazilian Amazon in a National Context: Economics, Policy, and Practice. Society & Natural Resources 10 (5): 433-451

Brazilian Institute of Geography and Statistics (IBGE). Brazilian Census 2000, www.ibge.gov.br/censo,

Budowiski G (1984) The Role of Tropical Forestry in Conservation and Rural Development. The Environmentalist 4(7):1573-2991

CARVALHO JL (1995) Sistemas Agroflorestais Como Alternativa Auto Sustentável Para o Estado de Rondônia. Histórico, Aspectos Agronômicos e Perspectivas de Mercado, EMBRAPA-EMATER

Caviglia JL, Kahn JR (2001) Diffusion of Sustainable Agriculture in the Brazilian Tropical Rain Forest: A Discrete Choice Analysis. Economic Development and Cultural Change 49(2):311-333

Executive Planning Commission for Cultivation of Cacau (CEPLAC). Retrieved from: http://www.ceplac.gov.br/download/download.htm

Current C, Lutz E, Scherr SJ (1995) The Costs and Benefits of Agroforestry to Farmers. World Bank Observer. 4(2):151-180

Chomitz KM, Thomas TS (2001) Geographic Patterns of Land Use Intensity the Brazilian Amazon. The World Bank, Development Research Group, Policy Research Working Paper 2687, Washington DC

Dubois JCL, Vianna VM, Anderson AB (1996) Manual Agroflorestal para a Amazonia. Vol. 1. Rio de Janeiro: REBRAF/Fundação FORD

Egler CAG (2001) Recent Changes in Land Use and Land Cover in Brazil.In: Daniel JH, Maurício TT (eds) Human Dimensions of Global Environmental Change. Brazilian Perspectives, Brazilian Academy of Science, Rio de Janeiro

Fearnside PM (1996) Agroforestry in Brazil's Amazonian Development Policy: Its Role and Limits as a Use for Degraded Lands. National Institute for Research in the Amazon (INPA), http://philip.inpa.gov.br/publ_livres/Preprints/1995/AGROSILV-eng.pdf

PM 1990. Predominant Land Uses in Brazilian Amazonia. National Institute for Research in the Amazon (INPA), http://philip.inpa.gov.br/publ_livres/Preprints/1990/PREDMC15-CUP.pdf

Fearnside PM, de L. Ferreira G (1984) Roads in Rondonia: Highway Construction and the farce of unprotected reserves in Brazil's Amazonian forest. Environmental Conservation 11(4):358-360

Fujisaka,S, White D (1998) Pasture or Permanent Crops after Slash-and-Burn Cultivation? Land-Use Choice in Three Amazon Colonies. Agroforestry Systems 42:45-49

Godfrey BJ, Browder JO (1997) Rainforest Cities: Urbanization, Development and Globalization of the Brazilian Amazon. Columbia University Press

Medrado MJS (1994) Sistemas Agroflorestais: Aspectos Técnicos Básicos e Indicações in Galvão, A. P. M . (Eds) Reflorestamento de Propriedades Rurais para Fins Produtivos e Ambientais: Um guia para ações municipais e regionais. Colombo: Embrapa Florestas.

National Institute for Space Research (INPE). Amazon Forest Deforestation Monitoring. http://www.obt.inpe.br/prodes/prodes_1988_2005.htm

Nepstad D, Uhl C, Pereira CA, Da Silva JMC (1996) A Comparative Study of Tree Establishment in Abandoned Pasture and Mature Forest of Eastern Amazonia. Nordic Ecological Society Oikos 76(1):25-39

Stickler CM, Almeida OT (2006) Globalization of the Amazon Soy and Beef Industries:Opportunities for Conservation.Conservation Biology 20(6):1595-1603

Kaimowitz D, Mertens B, S.Wunder S, Pacheco P (2004) Hamburger connection fuels Amazon destruction. Center for International Forestry Research, Bogor, Indonesia, www.cifor.cgiar.org

Nair PKR (1993) An introduction to Agroforestry. Kluwer Academic Publishers, Dordrecht, Netherlands.

Pedlowski MA, Dale VH, Matricardi EAT, da Silva Filho EP (1997) Patterns and impacts of deforestation in Rondônia, Brazil. Landscape and Urban Planning 38 (3):149-157

Pfaff ASP (1999) What Drives Deforestation in The Brazilian Amazon? Evidence from Satellite and Socioeconomic Data. Journal of Environmental Economics and Management 37(1):26-43

Walker R, Homma O (1996). Land use and land cover dynamics in the Brazilian Amazon: An overview. Ecological Economics 18(1):67-80

Rudel TKOT, Coomes E, Moran F, Achard A, Angelsen JC, Xu, Lambin E (2005) Forest transitions: towards a global understanding of land use change. Global Environmental Change:Human Policy Dimensions 15:23–31

Wilson EO (1999) The Diversity of Life. W.W. Norton and Company, Inc. New York

CHAPTER 13 - Agricultural Land-Use Trajectories in a Cocaine Source Region: Chapare, Bolivia.

Andrew Bradley[1] and Andrew Millington[2]

[1]*Centre for Ecology and Hydrology, Monks Wood, Abbots Ripton Huntingdon, , PE28 2LS, UK*
[2]*Department of Geography, Texas A&M University, College Station, TX 77843-3147, USA*

ABSTRACT

In the Amazon basin, forest clearance models describe the processes of land-cover conversion from forest to agriculture and offer an insight into trajectories of land use in new agricultural areas as farmers respond to changes in household demographics, biophysical parameters and economic conditions. At the foothills of the Andes in the Chapare of Bolivia, there are several colonization areas where coca leaf, the base ingredient for cocaine, is grown. The coca economy is illegal in these areas and has not been integrated into colonist clearance models. We have examined the impacts of this illicit economy in three communities in Chapare where colonists were interviewed about their farming activities since settlement began in 1963. Changes in cropping patterns have been linked to land-cover maps derived from satellite images. In one community, land-cover patterns became more complex as the illicit coca economy introduced a new set of drivers, different from those in the Amazon clearance models. The drivers, which were present at household, landscape and national levels, have acted in tandem at same periods in the past four decades whilst during other a single driver has dominated. Colonization, which began in the 1960s, became dominated by a lucrative illegal coca economy from the mid-1970s onwards, which was then gradually counteracted by anti-coca policies from the mid-1980s. The resulting land-cover trajectories were different than those described by colonist models, and land cover-change outcomes were different. Understanding these new drivers in an illicit economy is helpful towards directing policy on conservation and sustainable farming in the foothills of the highly diverse eastern Andean slopes.

INTRODUCTION

In the latter half of the 20th Century tropical forest loss in South America has largely been induced by economic development. Amazonian forests have been cleared through several Brazilian development programs (Rudel 2005, Walker and Hommer 1996) and forests have been converted to various forms of agriculture. Generally the Brazilian situation can be applied to other parts of the Amazon Basin (Achard et al. 1998, Eva et al. 2004), with Bolivia (e.g. Steininger et al. 2001) being the focus of this chapter. Several concerns arise as forests are converted to other land covers. The global carbon balance and the feedbacks on climate change may be modified (Foody et al. 1996, Houghton et al. 2000, Naughton-Treves 2004), a loss of forest habitat impacts biodiversity and connectivity (Myers et al. 2000; Skole and Tucker 1993) and agricultural encroachment into forests causes forest fires to become more widespread (Cochrane et al. 1999; Cochrane 2003). Further forest loss through the continuation of development projects (Laurence et al. 2001) indicates that there is a pressing need to understand these agricultural processes. By deepening our understanding of the drivers of the forest-to-agriculture conversion, policymakers can help mitigate these environmental impacts through the development of better informed policies.

Not surprisingly the varying intensities of landscape transformation found within political units cannot be explained alone by using macro-level processes (McCracken et al. 2002) because land-cover change processes operate below the national level. For example, at the landscape level, soils and topography may influence the farming choices (Brondizio et al. 2002). Some drivers cause forest to be replaced with agriculture, a further set operate at the household level on the new agricultural lands that have been opened up through colonization and may or may not influence decisions to clear more forest. To mitigate the environmental consequences in areas of colonization all of the drivers and land-cover change processes need to be identified down to the household level.

At the household level, current concepts of colonist agriculture describe how colonists first clear forest to create a few hectares of land for subsistence. After a period of consolidation more forest is cleared for a perennial crop, or to create pasture for livestock, depending on labor, credit and subsidies at the household level. When there is demographic change in the household, the farm labor force available to maintain a farm can be increased and the balance between pasture and perennials begins to vary. Conversely, the labor force can also decrease and as farming activities are abandoned forest regrowth may occur. Household and regional level models have been proposed to describe the drivers of colonist agriculture, (e.g., Walker 2004; McCracken et al. 1999) and rates of forest clearance in Amazonia (Brondízio et al. 2002). These models have also described typical land-use and cover change trajectories in deforested areas.

Illicit activities in colonist areas, such as illegal coca leaf cultivation, the base ingredient for cocaine, also impact household decision making (Bradley 2005). Little is known about the impacts illicit cultivation has on forest consumption, land-use decision making, land-cover change trajectories and whether the consequences are different from current descriptions of colonist clearance. We address these unknowns in this chapter. As all the coca used to produce cocaine is grown along the eastern slopes of the Andes - a zone of high biodiversity globally (Myers et al. 2000 – analysis of such a region is potentially more useful in conventional terms than those from in other, less diverse parts of the Amazon Basin.

In this chapter we identify the main processes causing agricultural change in the Chapare of Bolivia. A brief overview of data collection is followed by the description of land-use and land-cover drivers. Their impacts on land-cover change trajectories are compared to land-cover change trajectories for other colonization zones in the Amazon.

STUDY AREA AND BACKGROUND

Chapare is in the Cochabamba Department of Bolivia, located at the foot of the Andean mountain chain. The area, centered on 17.00° S, 65.00°W, covers approximately 6000 km^2 and was originally covered with lowland humid tropical forests grading into montane forest to the south and inundated humid forest to the north (Montes de Oca 1989). Since the early 20th Century, the area has seen small sporadic attempts at colonization (Rodríguez 1997) which did not have any serious threat to the forest cover until the Bolivian government implemented economic plans to diversify an economy dependent on mineral extraction whilst encouraging import substitution (Kaimowitz et al. 1999; Pacheco 1998, 2002). In 1961 Chapare was designated as one of three areas totaling one million hectares for agricultural expansion in Bolivia's lowlands. Government-sponsored colonists were organized into separate farming communities, normally containing around 100 families, each being allocated approximately 20 hectares of land for cultivation (Henkel 1971, 1973), or 50 hectares if it was designated as a pastoral community. In the 1980s the construction of a new road between the cities of Cochabamba and Santa Cruz to bypass the existing tortuous mountain route improved in Chapare and colonization accelerated.

In the 1970s a small scale coca and cocaine economy existed in the region. Two events coincided in the 1980s which increased significantly the area under coca cultivation. First, in 1986 the government reorganized the tin mining industry leaving 23,000 miners unemployed. Secondly, there was a major increase in the demand for cocaine globally. Hence a large labor force became available and impoverished migrants from the highlands (not just out of work mining families) moved to Chapare and became involved in cultivating, processing and trading coca (Rivera 1990). In the 1980s and 1990s, the 'war on drugs' implemented by the United States government (Morales 2004) saw a number of anti-coca policies focused on Chapare to eliminate the base ingredient of cocaine. Polices were enforced to eradicate coca bushes and alternative development programs were created to promote and encourage legal replacements for coca bushes (Léons and Sanabria 1997). By 2002 the Bolivian government announced that coca was eradicated in Chapare, and farmers were growing alternative crops and rearing livestock. Consequently deforestation was well entrenched between 1986 and 2000, and by the start of the current decade fragmentation of the forest estate in Chapare was well advanced and had been replaced with various forms of colonist agriculture (Millington et al. 2003).

METHODS

During a wider study of land-use and land-cover change in Chapare (Bradley 2005, Bradley and Millington, 2008), methods based on the approach of the IHDP-LUCC programme (Rindfuss et al. 2003) using image analysis, social science participatory research methods, and the analysis of economic data were used to study three communities: Arequipa, Bogotá, and Caracas[1].

Cloud-free imagery acquired during the dry season (May to October) was selected from MSS, TM and ETM image archives. Five image time points were used each community[2]. After pre-processing, land-cover maps (in a forest and non-forest format) were created using the Erdas ISODATA

[1] The names of the communities have been changed because coca has been grown illegally in these communities in the past.

[2] Imagery from 1976, 1986, 1993, 1996 and 2000 were obtained for Arequipa and Caracas; and from 1975, 1986, 1992, 1996 and 2000 for Bogotá which is located on an adjacent Landsat scene (Bradley, 2005).

clustering algorithm (Erdas 2001). A property grid for each community, provided by the Instituto Nacional de Colonisation (INC) was superimposed on each satellite image (Bradley 2005).

The communities were visited to verify the land-cover classes and discuss land management with farmers. After being granted permission to work in each community from the leaders of the *sindicatos*, volunteers were recruited through participation and attendance of community agricultural meetings. Participatory research included questionnaires aimed at verifying current and past land-uses and the reasons for changes in land management and crop switches; walking tours of farms to map and discuss land-use practices (using field sketches and GPS survey); and discussions which centered on the current and past land-cover maps to establish the land-use histories of farms. Communities were visited in 2002 and 2003 and interviewees represented 22 percent, 18 percent and 13 percent of households in Arequipa, Bogotá, and Caracas respectively.

The results of the image analysis were used to describe the changes in agricultural land cover and the results of the interviews were used to describe the temporal land-use patterns and the drivers of agricultural land-use change.

RESULTS

The Land-Use Activities and Land-Cover Patterns, 1960-2003

Temporal changes in the major land-management activities are shown in Figure 13.1. Arequipa was settled in 1983 and farmers initially cultivated mainly coca bushes, but by the mid 1990s cultivation was expanded to include bananas. By the late 1990s many farmers had started to cultivate coca substitute crops other than bananas (e.g., black pepper, heart of palm and passion fruit). In Bogotá, a few farmers settled in the early 1970s to cultivate rice, but the majority settled in 1983 to raise livestock. Forest was cleared to create pasture, though rice was always cultivated before turning the land over the pasture. Coca was often cultivated on farms in Bogotá in the 1980s. By the mid-1990s, cattle-rearing was the main activity, though a few farmers additionally cultivated coca and substitute crops. In Caracas, which was settled from 1963 onwards, farmers initially cultivated rice and citrus. By the late 1970s and through the 1980s, coca and citrus came to dominate land use; but by the 1990s farmers had to turned their attention to citrus, bananas and coca substitute crops. In all three communities, subsistence was maintained throughout to ensure food security for each household.

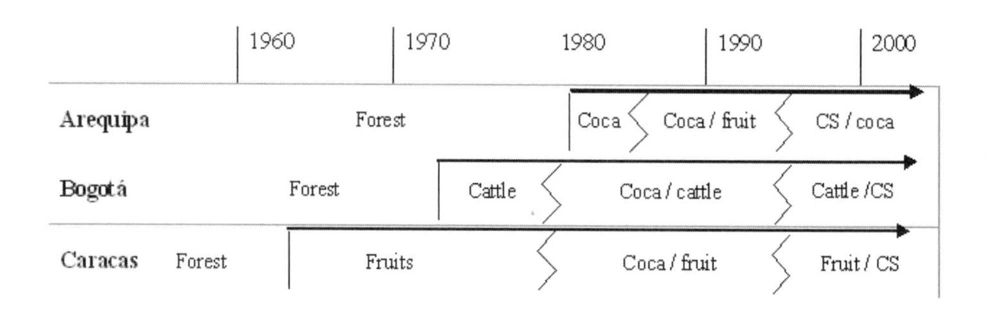

Fig. 13.1. A comparison of the major land-management activities in the three communities between 1963 and 2003 (Source: Field interviews). Straight, vertical lines indicate when the communities were settled, and the zig-zag lines indicate a transitional period between the changes in major land-use. CS refers to coca substitutes.

A cross comparison among the three communities provides three main observations. First, each community had their own distinct land-management pattern. But whilst there were distinct land-management patterns, agriculture in all three communities was dominated by coca cultivation at some time from in late 1970s to the early 1990s. Third, coca cultivation became less important and farmers reverted back to their initial activities or became involved in the cultivation of coca substitute crops from the early 1990s.

Figure 13.2 illustrates maps of land cover for one community (Arequipa) for four time points between 1986 and 2000. A property grid is superimposed on each image so individual land-cover changes can be seen for each farm. In the image sequence, forest area (dark gray[3]) decreased as clearance occurred and a variety of crops were planted (light gray[4]). The number of non-forest classes representing crops and fallow regrowth increased from three in 1986 to five in 2000.

Farmers were interviewed about changes in the land-cover classes using a sequence of forest/non-forest maps created for their plots from imagery, so the plot histories could be developed and land-management decisions interpreted. In 1986 the small number of classes and contiguous areas of each class reflected synchronicity between land-management decisions of the farmers as they cleared forest for subsistence and coca. During the

[3] Green in the on-line version

[4] Blue, brown and yellow tones in the on-line version

1980s land-management activities broadened as more farmers began to cultivate bananas, either as a replacement or as a supplement to illegal coca. In the 1990s diversification continued as more coca substitutes were encouraged. The increase in the complexity of land-use patterns over time reflects the fact that individual farmer's land-management decisions diverged.

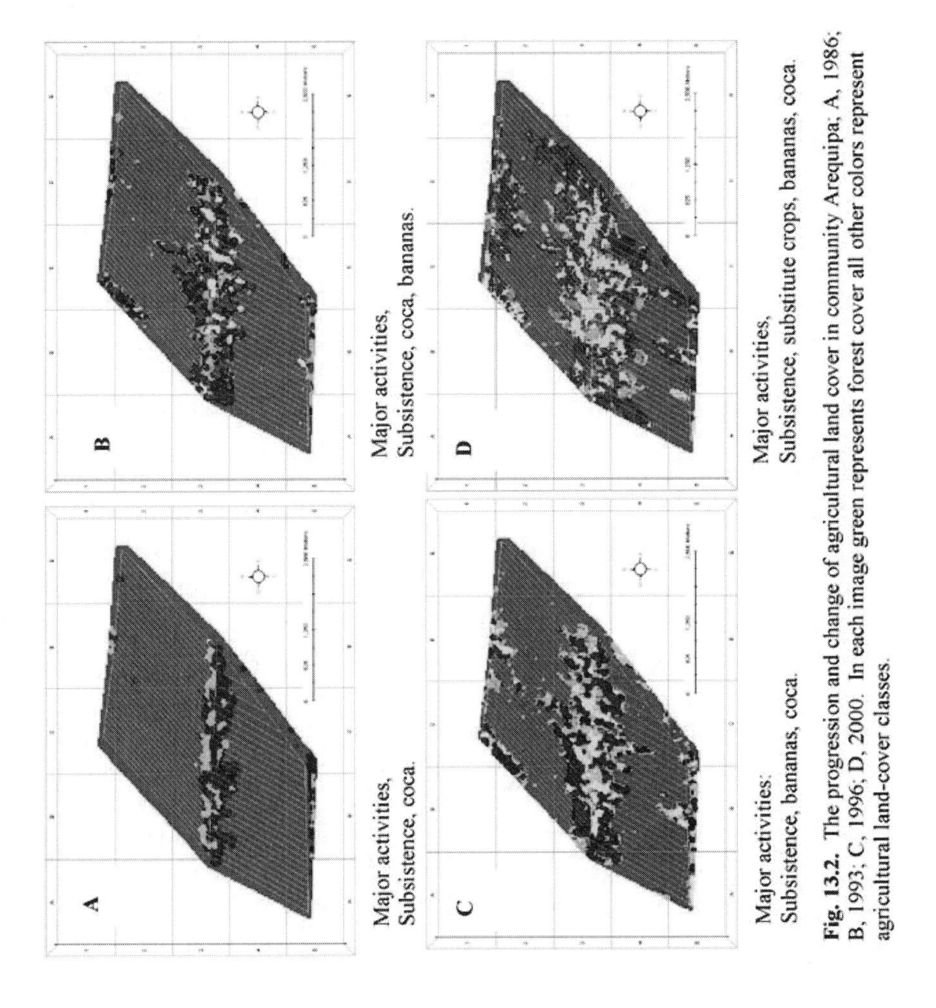

Fig. 13.2. The progression and change of agricultural land cover in community Arequipa; A, 1986; B, 1993; C, 1996; D, 2000. In each image green represents forest cover all other colors represent agricultural land-cover classes.

The Drivers of Land-Use 1960-2003

Explaining the causes of land-use changes in expanding agricultural areas requires an understanding of land-use change drivers. To do this we asked farmers to recall when and why they changed crops on their farms. We compared these observations with statistical data on areas of coca bushes and substitute crops (from government and USAID development programs) and price data for coca and coca substitutes for the whole of Chapare.

Trends in the cultivated areas of coca and substitute crops (Figure 13.3) for the region show that the land-management strategies adopted by the farmers we interviewed were generally replicated across the entire colonization zone. Coca cultivation[5] in Chapare increased from 10,000 hectares in 1980 to over 40,000 hectares in 1989 and then declined to just over 30,000 hectares in 1991. The 1980s peak in cultivation corresponds to the time period when our farmers recalled cultivation of coca in their communities generally, as well as on their farms specifically. The switch to coca began in the 1970s after a decade of weak government support associated with failed development policies. Farmers experienced difficulty moving bulky, perishable goods out of the area - roads were in poor condition and seasonally impassable. Gradually more and more farmers reverted to the illegal cultivation of coca as dried leaves could be transported in sacks slung over the shoulder without relying on motorized transport (Riviera 1990). In the early 1980s a global demand for coca ensured exceptionally high prices for the leaves (Bostwick et al. 1990), and this provided a further incentive for many farmers to switch to illegally cultivated coca. As a consequence the area under cultivation increased between 1980 and 1989 (Figure 13.3).

Pre-existing farming activities were affected by the boom in coca cultivation in the following ways. First, communities established by colonists in the 1950s and 1960s began to concentrate on illegal coca and rely less on or even abandon legal crops. This was the case in the oldest community we studied - Caracas - where citrus and rice became secondary to coca. Secondly, in communities settled during the coca boom in the 1980s (e.g., Arequipa), colonists cultivated illegal coca on their new cleared

[5] Coca was traditionally cultivated in the sub-tropical valleys of the eastern slopes of the Andes (*yungas*) in Bolivia. One element of the anti-narcotics policies that have been institgated is law 1008 passed in 1988 which allows coca to be legally cultivated in traditional coca growing areas in the *yungas* of Cochabamba and La Paz Departments, thereby unambiguously making coca grown in the lowlands of Chapare illegal.

plots. Very few farmers cultivated a secondary crop such as bananas. Thirdly, in cattle rearing communities such as Caracas, farmers continued to expand livestock operations, but also often supplemented their income by cultivating coca.

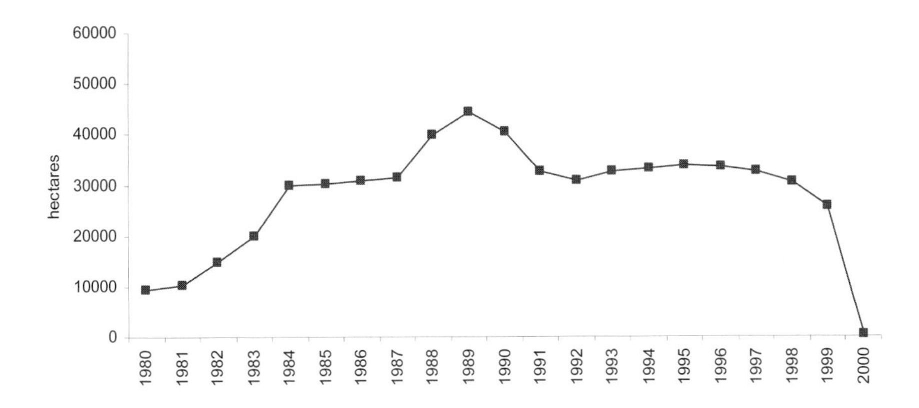

Fig. 13.3. The area under coca cultivated in Chapare between 1980 and 2000 (CEDIB 2000). Coca cultivation peaked ay 44,300 ha in 1989, the same time it became the main crop in Arequipa, Bogotá and Caracas. Propoganda associated with government targets are likely to have influenced the zero value in 2000, rather than actual elimination of the crop.

Therefore during the coca boom, farmers concentrated their efforts on small areas where they cultivated coca and paid less attention to crops they had grown previously. These changes were partly driven by the profitability of the coca leaf which gives much higher returns per hectare than other crops (Table 13.1). Others factors supported this main driver of the switch to coca. First, coca leaf had a reliable, stable market in comparison to substitute crops. For example, the price of bananas, a typical alternative to coca, was quite variable during the 1990s (Figure 13.4). Over a similar same time period the price of a 45 kg sack of coca leaves remained stable at approximately 50 USD (Institute National de Estadisticas, Bolivia). Uncertainties in the markets of coca substitutes can also be seen in price fluctuations for oranges, mandarins and pineapples (CORDEP/DAI 1993, 1999). Given the free market economic conditions that prevailed, coca leaf was the obvious economic option for most farmers during the 1980s.

Table 13.1 Crop profitability of one hectare land cleared for difference crops in Chapare, 1991.

Crop	Profit (USD/ha) in 1991	Land index
Coca	17,714	1
Pineapple	10,022	1.1
Black pepper	5,162	2.2
Heart of palm	4,782	2.4
Orange	4,280	2.7
Banana	2,622	4.3

The land index indicates the hectares required to equal the profit from a hectare of coca, assuming a price of 75USD for 45kg of coca leaves. These values have varied since 1991, yet illegal coca has remained more profitable than any of the legal crops. (Source: DOA 1991)

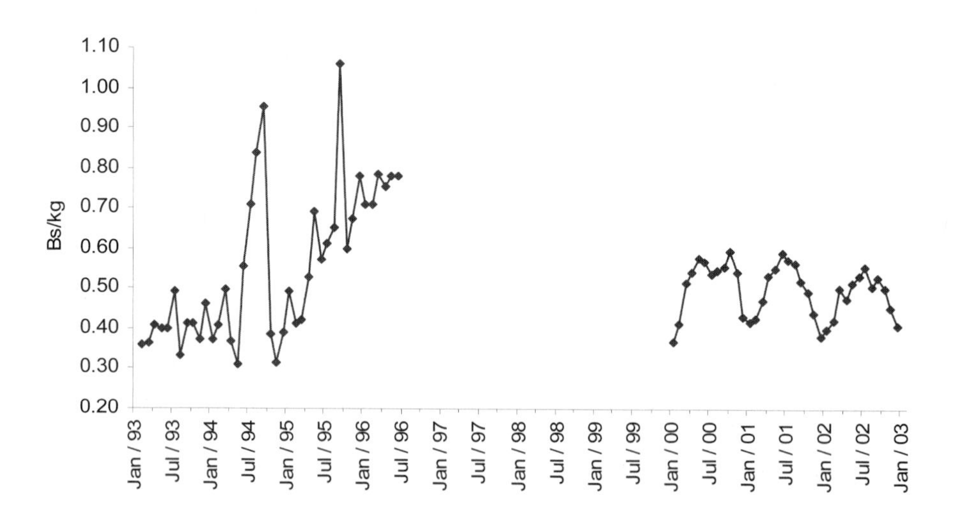

Fig. 13.4. Wholesale banana prices in Bolivian bolivianos per kilogram between January 1993 and June 1996 (CORDEP/DAI, 1993, 1999) show increasing but npredictable prices of bananas, and between 2000 and 2003 (Proyecto CONCADE, USAID, 2003) when subsidies maintained a stable seasonal price. No data were available between June 1996 and January 2000.

In the late 1980s, land-management practices changed again in Chapare despite the continued economic benefits of cultivating coca. The area of illegal coca cultivation began to fall after 1989 (Figure 13.3) and from 1986 the area of legal coca substitute crops increased from just under 40,000 hectares to almost 120,000 hectares in 2000 (Figure 13.5).

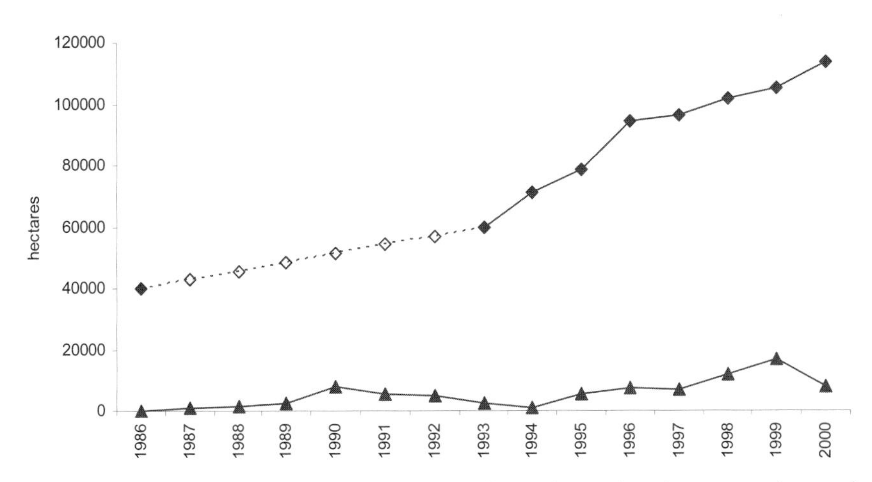

Fig. 13.5. The area of substitute crops and eradicated coca in Chapare. Diamonds show the rise in area of substitute crops between 1986 and 2000 (CIDRE, 2000). There were no data before 1986 or between 1987 and 1993: dotted lines indicate data interpolation. Triangles show the area of eradicated coca between 1986 and 2000 (UNODC, 2004)

The farming activities in the three communities surveyed indicated that by the beginning of the 1990s farmers were less likely to cultivate illegal coca, and by the end of the decade more farmers had been encouraged to grow legal alternatives by choice with the help of substitute programs or had returned to their old legal activities. By the mid and late 1990s, a gradual strengthening of anti-coca policies, including the increase in area of eradicated of coca bushes since 1986, had influenced land management (Figure 13.5). Alternative development programs encouraged farmers with subsidies that would help stabilize the prices of substitute crops. For example, from 2000 to 2003 banana prices had fallen from 1996-levels to approximately 0.5Bolivianos/kg, yet there was a stable seasonal pattern in prices in comparison to the wilder fluctuations experienced between 1993 and 1996 (Figure 13.4).

Our evidence shows that land-management drivers changed over time. Weak national development policies were eventually overtaken by the economic advantage of coca cultivation. This in turn was transcended by the increasing momentum of anti-coca policies. The outcome of these changes in drivers was a complex set of land-use changes within the agricultural area as illustrated in Figure 13.2. Farmers responded to the drivers at different times to other households. They did not always select the same substitute crop, and each substitute required different areas of land to match the returns from coca (Table 13.1). Adding to the complexity of land-use patterns was the fact that as land-use trajectories changed crops were also eradicated, abandoned, or replaced.

APPLYING THE 'COLONIST MODEL' IN CHAPARE

We have identified distinct changes in land-management activities in the three communities (Figure 13.1), but did the illicit coca economy have any impact on the conceptualized ideas of colonist land-use trajectories? The models of colonist clearance described by McCracken (1999, 2002), Scatana et al. (1996) and Walker (2004) can be used as a benchmark for a comparison with the land-use trajectories in Chapare and termed the 'colonist model' in this chapter. Walker (2004) has already identified that in different locations of the Amazon there was an evolutionary dynamic in colonist agriculture. In each colonist household, a period of subsistence is followed by cultivation of perennials and/or pasture expansion after which the agricultural activities are divided between similar proportions of annuals and perennials. Similarly McCracken et al. (2002) described the evolution of colonist activities where farm land-use trajectories may follow one of two scenarios:

(i) five years of annual cash crops, followed by five years of pasture, after which the farm is consolidated with perennial crops; or
(ii) five years of cash crops, followed by five years of perennials or pasture and then the farm is consolidated with both perennials and pasture (the proportion of which depends on the circumstances of the household).

This model is progressive, and changes in household demography and capital may force labor and capital on agriculture activities. As a result of these constraints, deforestation decreases and secondary succession increases (McCracken et al. 1999).

In the three Chapare communities, it was clear that all farms began with land-management orientated around subsistence, and depending on the community after two to three years they entered an economy based on fruit crops, coca or rice and pasture, thereby adhering to the concept of the colonist model. Progression toward consolidation of crops may have occurred in Caracas, but this cycle was disrupted in the 1980s when the strong coca economy influenced the land-management decisions of farmers, and again later in that decade when anti-coca policies further influenced the land-use decisions of farmers. The influence of household demography, landscape scale controls, and national level policies were reduced. Paradoxically farms could have been consolidated with a coca crop. This would have adhered to the colonist model but they did not because of the actions of the Bolivian government which embarked on policies to destroy or offer compensation for the voluntary removal of coca crop and encourage substitutes through alternative development programs that had short-term subsidies or no support at all.

The impact was that predictable, long-term land-use trajectories were not maintained. Furthermore, in the presence of a lucrative crop with a stable market, as opposed to the uncertain markets of poor paying alternative crops, the temptation of farmers to continue or return to the cultivation of illegal coca was too great and this further encouraged land-use decisions-making based on short-term phenomena. The interplay of these decisions across the Chapare landscape have created a regional land-management scenario that is more complex than the crop/pasture dynamics described for other Amazonian colonist areas. Most notably, the enforcement of anti-coca policies has impacted the 'colonist model' land-use trajectories. Table 13.2 shows how land-use trajectories have progressed in relation to changes in drivers in Chapare.

In the 1960s, the drivers of colonist agriculture were the government sponsorship schemes mainly promoting the cultivation of perennial crops or cattle rearing. After subsistence the land-cover trajectory went from forest to legal crops or pasture. These colonists were pioneers battling with the environmental conditions of the humid tropical forest[6]. Following a succession of failed crops farmers realized that rice cultivation and fruit cultivation were the best options. Therefore, in the 1960s and 1970s, household decisions followed those which were typical of the early stages in the colonist clearance cycle elsewhere in the Amazon Basin, and the land-use trajectories adhered to the idea that colonists moved from subsistence to perennial cultivation (Table 13.2).

[6] Most of the colonists in Chapare come from the Bolivian arid and semi-arid highlands and initially lacked farming experience in the lowland, humid tropics.

Table 13.2. Conceptualisation of land use trajectories in Chapare from the 1960 to 2000. The duration of the dominant drivers: national policy; the coca economy and; anti-coca policies are shown. Solid lines indicate the periods of dominance. Over time the trajectories become more numerous and complex in response to the changing drivers.

Trajectory		change	1960	1970	1980	1990	2000
forest	to	fruits, pasture					
forest	to	coca					
coca	to	crops and / or pasture					
crops and / or pasture	to	subsidised crops and / or pasture					
crops and / or pasture	to	intercropped coca					
any land use	to	secondary succession					
secondary succession	to	any land use					
Drivers of land management		National development policies					
		The coca economy					
		Anti-coca policies					

Beginning in the 1970s and through the 1980s, the high price of coca and accessibility of a market for this crop caused farmers to cultivate coca immediately after clearing forest or replacing crops and pasture in favor of coca bushes (Table 13.2). In the 1980s and 1990s, land-use trajectories became even more complex with the enforcement of anti-coca policies. Eradication, voluntary substitution, and alternative development programs forced more and more farmers to replace coca with substitute crops or pasture. Land-use trajectories were further influenced by the wide choice of fruit crops, each with different economic benefits and land requirements. The choices of substitute crops could not always be sustained because of the economic competitiveness of coca (Table 13.1). Additionally, because many farmers in Chapare changed over to substitute crops, oversupply in these markets caused price instabilities which were manifest in boom–and-bust cycles. With demand lacking and occasional absence of a market for substitutes (Hellin 2001), farmers would cultivate more coca or revert to a different substitute. Interviews revealed that the complexity of land-use trajectories was complicated by the fact that a coca substitute was limited to the lifetime of the crop or time over which a subsidy was available. Farmers would then change to a different crop if a subsidy for that became available.

Farmers could not sustain a perennial crop as suggested should happen in the colonist models because the measures to counter coca cultivation did not motivate a household to consolidate farming around a single perennial crop. To avoid detection, in 2000 and 2003 we observed coca inter-cropped with fruit bushes or other crops to avoid detection of coca bushes by the authorities and in doing so have created an additional land-use trajectory (Table 13.2). Indications are that this strategy had become more widespread in the 1990s to avoid aerial surveys.

Even though the drivers of land-management activities were dominated by coca and anti-coca policies, some processes of land change identified in the colonist model were still operational. The colonist model eventually progresses to a point where secondary succession increases and deforestation slows down. In Chapare, Millington et al. (2003) described how forest regrowth was on the increase in some areas by the end of the 1990s. This was confirmed by farmers in the three Chapare communities who described that they had abandoned crops and allowed forest to grow where crops or pasture had existed previously. In addition, the colonist model explains secondary succession as a result of changing household labor and capital characteristics. In Chapare, some farmers attributed forest regrowth to labor constraints which arose when their children had moved to new plots or to nearby cities.

By the late 1980s and 1990s, other help dynamics explain land-change in the Chapare. Increasing numbers of farmers in the older communities had cleared their full quota of 20 hectares, and in these cases, farmers had no option but allow a fallow period of forest regrowth in order to recuperate soils. Forest regrowth could also be initiated by the abandonment of substitute crops when subsidies ceased. As the anti-coca policies gained strength and encouraged the cultivation of coca substitutes, farmers described how they diversified the number of substitutes in order to access more than one market and decrease their risks by diversification. In the communities visited, farmers concentrating on fruit cultivation seldom incorporated pasture with the cultivation of crops, as described in the colonist models and those that had livestock may only have had a few cows, because the allocation of 20 hectares lots was too small to sustain large herds of livestock. In fact it appeared that if livestock was the major activity, this had been a conscious decision of the farmer from the beginning of settlement, community Bogotá being a case in point.

The evidence shows that the coca economy and anti-coca policies masked the land-use trajectories described in the colonist model and strongly influenced the land-management activities and land-use trajectories in Chapare. Such an impact occurred because the drivers dominated at national, landscape, and household scales. At the national level, macroeconomic policies encouraging agricultural productivity and colonization process during the 1960s and 1980s were overcome by the strong illicit coca market. Anti-coca policies, main the land-change drivers in the 1980s and 1990s, were not based on developing Chapare as an economic region of Bolivia, but were a political decision to satisfy the demands of the US government in order to secure development aid. At the landscape level, farming activities were not just a function of topography and soils. In the 1960s land-management choices were driven by experimentation until farmers learned how to cope with their new environments. From the 1970s onwards coca became the crop of economic choice and could be cultivated anywhere in Chapare regardless of soil. In the 1980s, encouraging farmers to grow alternatives to coca was a much higher priority than identifying suitable areas and soils for particular crops which did not become a concern until the 1990s. At the household level, the predicted patterns and drivers of land-use trajectories were not just a function of demographic change, available credit, and investment capital because households were influenced by the reliable, lucrative illegal coca market. These activities were most dominant in the late 1970s and 1980s. In the 1980s and 1990s the anti-coca policies were implemented, forcing farmers to uproot coca bushes and grow substitutes, regardless of family size or available capital.

THE IMPACTS OF THE ILLICIT COCA ECONOMY

This chapter has examined the effects the illicit coca economy has had on forest consumption, land-use decision making, land-cover change trajectories in Chapare and tried to evaluate whether the consequences are different from current descriptions of colonist clearance elsewhere in Amazonia. The three communities that have been studied have shown that the illicit coca economy has influenced the activities of colonist farmers. Crucially, the consequences of an illicit economy have not been identified in other Amazonian colonist models. Specifically, the land-cover change trajectories became distorted and no longer followed the patterns predicted by the colonist model. In the 1960s household demographies, capital constraints, and a lack of local environmental knowledge ensured that farmers followed similar land-use dynamics to other Amazonian colonization areas. In the 1970s and 1980s, the coca economy dominated and farmers were drawn away from the perennial/pasture land-use trajectory seen elsewhere due to the emergence of a lucrative perennial which had a secure market. Households found that the strength of the coca economy sustained farming much better than the policies and markets supporting the original colonization.

During the 1980s and 1990s, the intervention of strong, forceful anti-coca policies continued to divert events in Chapare away from the path of the colonist model. When anti-coca policies forced farmers to become reliant on crops supported by subsidies, or when farmers grew crops which had unstable markets, the land-cover change trajectories became complex. The enforcement of anti-coca policies perpetuated unsustainable agriculture at the expense of forest, as seen on the satellite imagery where (i) there was a continued loss of forest, and (ii) the pattern of land-cover change indicated almost continual switching of land-management activities.

With these impacts in mind, we have developed a better understanding of colonist agriculture in regions where, and at times when, illicit economic activities and policies to counter them are in place. By identifying the new drivers and their impacts on land-use and land-cover change trajectories we could meaningfully consider preventative measures for environmental protection in other coca areas, such as the Bolivian *yungas*, the upper Huallaga valley in Peru, and the eastern Andean foothills of Colombia. In areas where coca cultivation could dominate, such as northeast Ecuador or areas of forest yet to be colonized in coca-growing regions, policies must ensure that existing markets for the current agricultural activities are supported in times of economic weakness, otherwise coca could encroach.

ACKNOWLEDGEMENTS

Funding for this research came from the European Union (ERBIC189CT80299) which also contributed to Andrew Bradley's PhD along with a stipend from the University of Leicester, and the 2001 Slawson Award from the Royal Geographical Society with the Institute of British Geographers. We are grateful to Felix Huanca Viraca for accompanying us in the field and support from Universidad Mayor de San Símon, Centro de Biodiversidad y Genetica, Cochabamaba,

REFERENCES

Achard F, Eva H, Glinni A, Mayaux P, Richards T, Stibig H J (1998) Identification of deforestation hot spot areas in the humid tropics. TREES Series B, Research Report Number 4, Joint Research Centre, European Commission, Ispra, Italy.

Bradley AV (2005) Land-use and land-cover change in the Chapare region of Bolivia, PhD Thesis, University of Leicester, UK

Bostwick D, Dossey J, Jones J, Arrueta JA, Calla C, Laserna R (1990) Evaluation of the Chapare Regional Development project (CDRP), Pragma Corporation, Arlington, Virginia., Cochabamba, Bolivia.

Bradley AV and Millington AC (2008) Coca and colonists: quantifying and explaining forest clearance under coca and anti-narcotics regimes. Ecology and Society (in press).

Brondízio ES, McCracken SD, Moran EF, Siqueira AD, Nelson RD, Rodriguez-Pedraza C (2002) The colonist footprint. Towards a conceptual framework of land use and deforestation trajectories among small farmers in the Amazonian frontier. In: Wood CH, Porro R (eds) Deforestation and Land Use in the Amazon .University of Florida Press, Gainsville, Florida,

CEDIB (2000) Data on the area of coca in Chapare. Centro de Documentacion e informacion de Bolivia. Cochabamba, Bolivia.

Cochrane MA, Alencar A, Schulze MD, Souza Jr CM, Napstad DC, Lefebvre P, Davidson EA (1999) Positive feedbacks in the fire dynamic of closed canopy tropical forests. Science 284:1832-1835.

Cochrane MA (2003) Fire Science for rainforests. Science 421: 913-919

CONCADE/USAID (2003) Pricios myoristas de los principlaes mercados internationales gestion 2000. CONCADE (Counter-Narcotics Consolidation of Alternative Development Efforts)/USAID, Cochabamba, Bolivia.

CORDEP/DAI (1993) Guía informativa: banano, piña, palmito, maracuya. Proyecto de Desarollo Regional de Cochabamba (CORDEP), Cochabamba

CORDEP/DAI (1999) Guía informativa: banano, piña, palmito, maracuya, pimienta. Proyecto de Desarollo Regional de Cochabamba (CORDEP), Cochabamba

Erdas (2001) ERDAS Imagine Tour Guides v8.5. ERDAS, Atlanta

Eva HD, Belward AS, De Miranda EE, Di Bella CM, Gond V, Huber O, Jones S, Sgrenzaroli M, Fritz S (2004) A land cover map of South America. Global Change Biology 10:731-744

Foody GM, Palubinskas G, Lucas RM, Curran PJ, Honzak M (1996) Identifying terrestrial carbon sinks: classification of successional stages in regenerating tropical forest from Landsat TM data. Remote Sensing of Environment 55: 205-106

Hellin J (2001) Coca eradication in the Andes: Lessons from Bolivia. Capitalism Nature Socialism 12(2): 139-157

Henkel R (1971) The Chapare of Bolivia: a study of tropical agriculture. PhD thesis, University of Wisconsin, Madison, Wisconsin

Henkel R (1973) The Chapare Project:A study of directed colonization in the Bolivian tropics. Technical report, Arizona State University (unpublished)

Houghton RA, Skole DL, Nobre CA, Hackler JL, Lawrence KT, Chomentowski, WH (2000) Annual fluxes or carbon from deforestation and regrowth in the Brazilian Amazon. Nature 403:301-304

Kaimowitz, D.J. (1997) Factors determining low deforestation: The Brazilian Amazon. Ambio, 26, 536-40.

Laurance WF, Cochrane MA, Bergen S, Fearnside PM, Delamônica P, Barber C, D'Angelo S, Fernandes T (2001) The Future of the Brazillian Amazon. Science 291:438-439

Léons MB, Sanabria H (1997) Coca, Cocaine, and the Bolivian Reality. State University of New York, Albany.

McCracken SD, Brondízio ES, Nelson D, Moran EF, Siqueira AD, Rodriguez-Pedraza C (1999) Remote Sensing and GIS at farm property Level: demography and deforestation in the Brazillian Amazon. Photogrammetric Engineering and Remote Sensing 65(11):1311-1320

McCracken SD, Siqueira AD, Moran EF, Brondízio ES (2002) Landuse patterns on an agricultural frontier in Brazil. In:Wood CH, Porro R (eds) Deforestation and Land Use in the Amazon. University of Florida Press, Gainsville, Florida

Millington AC, Velez-Liendo XM, Bradley AV (2003) Scale dependence in multitemporal mapping of forest fragmentation in Bolivia: implications for explaining temporal trends in landscape ecology and applications to biodiversity conservation. ISPRS Journal of Photogrammetry and Remote Sensing 57(4):289-299

Montes de Oca I (1989) Geografia y Recursos Naturales de Bolivia. Academia Nacional de Ciencias de Bolivia, La Paz, Bolivia

Morales WQ (2004) A Brief History of Bolivia. Checkmark Books, New York

Myers N, Mitteermeier RA, Mittermeier CG, daFonseca GAB, Kent J (2000) Biodiversity hotspots for conservation priorities. Nature 403:853-858

Naughton-Treves L (2004) Deforestation and carbon emissions at tropical frontiers: A case study from the Peruvian Amazon. World Development 32(1): 173-190

Pacheco P (1998) Estilos de desarrollo, deforestación y degredación de los bosques en las tierras bajas de Bolivia. Centro de Información para Desarrollo (CID), La Paz, Bolivia

Pacheco P (2002) Deforestation and forest degradation in lowland Bolivia. In: Wood CH, Porro R (eds) Deforestation and Land Use in the Amazon. University of Florida Press, Gainsville, Florida

Rindfuss RR, Prasartkul P, Walsh SJ, Entwistle B, Sawangee Y, Vogler JB. (2003) Household-parcel linkages in Nang Rong, Thailand: challenges of large samples. In: Fox J, Rindfuss R, Walsh S, Mishra V (eds) People and the Environment: Approaches to Linking Household and Community Surveys to Remote Sensing and GIS. Kluwer, Boston

Rivera A (1990) Diagnostico socio económico de la población de Chapare. Central de Estudias de Realidad Economía y Social (CERES), Cochabamba, Bolivia.

Rodríguez G (1997) Historia del Trópico Cochabambino 1768 - 1972. Prefectura del Departmento de Cochabamba, Cochabamba, Bolivia

Rudel TK (2005) Tropical forests: regional paths of destruction and regeneration in the late twentieth century. Columbia University Press, New York

Scatana FN, Walker RT, Homma A, de Conto AJ, Ferreira CAP, Carvalho R, da Rocha A, dos Santos A, de Oliveira P (1996) Cropping and fallowing sequences of small farms in the "terra firme" landscape of the Brazilian Amazon: a case study from Santarem, Para. Ecological Economics 1:29-40

Skole DL, Tucker C (1993) Tropical deforestation and habitat fragmentation in the Amazon: satellite data from 1978 to 1988. Science 260:1905-1910

Steininger MK, Tucker CJ, Townshend JRG, Killeen TJ, Desch A, Bell V, Ersts P (2001) Tropical deforestation in the Bolivian Amazon. Environmental Conservation 28(2):127-134

UNODC (2004) Bolivia Coca Cultivation Survey. United Nations Office on Drugs and Crime. Vienna, Austria.

Walker R (2004) Mapping process to pattern in the landscape change of the Amazonian frontier. Annals of the Association of American Geographers 93(2): 376-398

Walker RT, Hommer A (1996) Land-use and land-cover dynamics in the Brazilian Amazon: An overview. Ecological Economics 18(1):67-80

Walsh SJ, Bilsborrow RE, McGreagor SJ, Frizelle BG, Messina JP, Pan WKT, Crewes-Meyer KA, Taff GM, Baquero F (2003) Integration of longitudinal surveys, remote sensing time series and spatial analyses: approaches for linking people and place. In: Fox J, Rindfuss R, Walsh S, Mishra V (eds) People and the Environment: Approaches to Linking Household and Community Surveys to Remote Sensing and GIS. Kluwer, Boston.

CHAPTER 14 - The Tobacco Industry in Malawi: A Globalized Driver of Local Land Change

Helmut Geist[1], Marty Otañez[2], and John Kapito[3]

[1] *Department of Geography & Environment, School of Geosciences, University of Aberdeen, AB24 3UF, UK*
[2] *Center for Tobacco Control Research and Education, University of California-San Francisco, San Francisco, CA 94143, USA*
[3] *Malawian National Human Rights Commission, Blantyre, Malawi*

ABSTRACT

The tobacco-dependent Republic of Malawi, formerly Nyasaland, is taken as an example of how the globalization of non-food cash crops such as tobacco impact upon land cover, in particular native, old-growth forests of the *miombo* biome. It is shown that markets and institutions harnessed by a handful of US leaf and cigarette companies, through a complex interlinking network of actors, account for a considerable share of local land change in the form of both land cover conversions and land quality modifications. Starting in late 19th century, tobacco monocultures became to dominate the whole country. Our insights run contrary to statements propagated by the tobacco industry and its agricultural front organizations. They also pose a major challenge for the implementation of the World Health Organization's first public health treaty to regulate use and production of tobacco.

INTRODUCTION

Land use is a highly political activity that requires an understanding of key actors and local situations for the design of effective policy interventions (Lambin and Geist 2006). We adopt a crop-specific approach to trace how the cultivation, processing and export of tobacco, a major colonial cash crop and the most widespread non-food crop worldwide, has transformed environments, economies and societies (Ponting 1992). The focus is on Malawi, a landlocked country in South-eastern Africa, which is one out of the 130 developing and transitioning countries where nearly 90 percent of global tobacco is grown. Malawi derives more than 70 percent of its foreign earnings from tobacco, making it the world's most tobacco-dependent economy (FAO 2003). It is the twelfth largest tobacco growing country (FAO 2006) and the eleventh largest tobacco exporter in the world (ITC 2006a). Transnational tobacco manufacturers such as British American Tobacco (BAT), Philip Morris (PM)/Altria, and Japan Tobacco (JT) use the majority of Malawian tobacco in cigarettes (Cox 2003, Geist 2006).

International development and aid agencies, in alliance with the global tobacco industry and its agricultural lobby, claim that tobacco cultivation – with special regard to central and southern Africa (ITGA 1992) – is a driver of economic development; hence the term "green gold" (UNCTAD 1995, Reemtsma 1995, FAO 1998). Today the spread of the tobacco epidemic is considered to be a global problem. The World Health Organization, through the Framework Convention on Tobacco Control (FCTC), established provisions to regulate tobacco use and production. It is acknowledged in the convention that a public policy issue arises which is linked to the design of tobacco as a smoke product with high nicotine enrichment and optimal combustibility[1]: agricultural practices such as topping and (de)suckering lead to rapid exhaustion of soil nutrients, making tobacco monocultures dependent on inputs of commercial fertilizers and biocides (alternatively, soil exhaustion is avoided by rotating crops or clearing forest land) (Goodland et al. 1984, Clay 2004); and, green tobacco leaves are dried up (cured) which requires substantial amounts of fuelwood that is usually taken from customary land (tobacco causes about five percent of deforestation annually in developing and transformation countries, amount-

[1] The FCTC, the world's first public health treaty, came into force in February 2005, and aimed to reduce tobacco use through measures such as calls on governments to ban tobacco advertising, put warning labels on packaging, and create smoke-free places. Also, article 17 in part IV (measures to regulate the supply of tobacco) asks for the "provision of support for economically viable alternative activities" for tobacco growers, while article 18 in part V (protection of the environment) addresses the need of "due regard to the protecion of the environment and the health of persons in relation to the environment in respect of tobacco cultivation" (http://www.fctc.org [accessed June 2006])

ing to as high as 26 percent in Malawi in 1990-95) (Clay 2004; Lambin and Geist 2006).

Claims continue that tobacco-driven land degradation and especially "deforestation associated with tobacco curing cannot currently be considered a significant negative externality" (ITGA 1995). The International Tobacco Growers Association (ITGA) delivered to the 1992 United Nations (UN) Conference on Environment and Development in Rio de Janeiro the message that tobacco's relative importance within the deforestation issue is minimal (Yach and Bettcher 2000), and the Tobacco Association of Malawi (TAMA) issued a press release in March 1996 that "tobacco crops are not responsible for deforestation".[2] Likewise, the Eliminate Child Labor in Tobacco Growing Foundation in Switzerland, an organization established in 2000 and funded by global tobacco companies, sponsors tree planting projects in Malawi. These efforts appear primarily aimed at replenishing the wood for curing tobacco and deflecting attention from tobacco-related deforestation, soil degradation, and water contamination in the country (Otañez et al. 2006).

In this chapter, evidence of land change in Malawi is provided that is linked to tobacco growing, negatively affecting land quality. We address the national scale, while specific details stem from growing areas in the Shire Highlands of southern Malawi. At the heart of our evidence are markets and institutions, following the working hypothesis that "[t]he influence harnessed by the tobacco industry [...] is planned centrally and extends globally through a complex interlinking network of players" (Yach and Bettcher 2000).

A HUMAN-ENVIRONMENT HISTORY OF TOBACCO

Prior to the British occupation in the early 1890s, local farmers in Nyɟasaland (now Malawi) were growing tobacco varieties for subsistence and trade, while commercial flue was first grown by European settlers in 1889 (Rubert 2005). Flue – also called Bright (Leaf) or (bright) Virginia – is cured in air-tight barns using artificial heat (mostly wood in Africa), which is distributed through a network of pipes or flues near the barn floor. For centuries, before the creation of flue, a diverse range of leaf types and curing methods were to be found internationally. Then, the catalyst for changing the global pattern of tobacco land

[2] http://www.tobaccoleaf.org/releases/malawood.htm V1 19 May 1996 [Online, June 2006].

use was the cigarette (especially in its white-stick, filter-tipped form), a tobacco product practically unknown before 1850. Two innovations came together in the creation of a cigarette: the invention machines that could automate the manufacture process and the accidental discovery (in New England) of a curing method using hot air and producing a bright-yellow leaf that was extremely well suited for use in cigarettes. Flue became the "global crop" which parallels the "global cigarette" (Geist 2006, Cox 2000).

After the 1899/1900 growing season, tobacco production in Nyasaland increased at ~75 percent annually (Rubert 2005), and by the 1920s it assumed a prominent place in the colonial economy and society (Kapito 2000). Following independence in 1964, Malawi found it difficult to escape from a tobacco-centred system which established manyfold "cultures of dependence" (Goodman 1993), "creating the Third World" (Ponting 1992) and affecting local (miombo) ecosystems (Misana et al. 1996).

Nyɟasaland was exploited for the benefit of the British home economy (in the 1990s, Malawian flue still serves United Kingdom [UK] markets, chiefly) (Table 14.1). The territory produced tobacco because the climate was suitable and cheap/unpaid labour available. Tobacco – 100-times more labor-intensive than wheat – resists large-scale mechanization and needs large amounts of intensive manual work over three to four months (currently, transnational tobacco companies in Malawi receive ca. US$10 million annually in economic benefits through the use of unpaid child labor). Tobacco took up much of the best land and largely displaced traditional subsistence agriculture. Local inhabitants were reduced to cultivating a small range of crops (primarily maize) grown on the poorest ground. Europeans inherited but disrupted traditional hand-cultivation systems which were well-adapted to the local environment, producing an agriculture that was stable, resilient, diverse, and capable of maintaining output over the long term. In contrast, tobacco monocultures triggered soil degradation (modification) and deforestation (conversion) as the most important types of land change (Ponting 1992, Clay 2004, Lambin and Geist 2006).

Through manifold institutional arrangements, local inhabitants were prevented from growing profitable cigarette tobaccos such as flue (until present) and burley (until the 1990s). Flue is grown on large-scale plantations owned and controlled by European settlers or members of the Malawian elite. With increasing exports to the United States (US), plantation agriculture transformed into large companies with an ever increasing share of transnational corporations such as Sable Farming Ltd. (McCracken 1983, Tobin and

Knausenberger 1998). In 2002, Universal Tobacco Corporation subsidiary Limbe Leaf in Malawi started contract growing with local farmers (Koester et al. 2004). Recently, Limbe Leaf started growing flue after taking control of the Kasungu Flue-Cured Tobacco Authority (KFCTA) and General Farms (Kadzandira et al. 2004). General Farms is a collection of over 60 large-scale tobacco farms formerly operated by Press Holdings and the Agricultural Marketing and Development Corporation (ADMARC).

TOBACCO MARKETS AND INSTITUTIONS

In post-colonial Malawi, tobacco continues to take a central role, with increasing tensions growing over the monopoly enjoyed by a handful transnational companies and increasing awareness about the need to explore alternative livelihoods for local growers. Since 1964, land under tobacco has been continuously extended well into the 21st century, with almost 10 percent of arable land (ca. 25 percent of the total land area) now devoted to tobacco (which is 16-times the global average).[3] Likewise, production levels increased steadily well into the 1990s (Figure 14.1). By 1994, Malawi gained the 10 million kilogram annual level of production, making the country one of the world's top producers of cigarette tobaccos. After Brazil, Malawi is the second largest producer of burley in the world in 2005 (Universal 2006), and it has long been among the top-15 flue exporting countries (Table 14.1).

Large segments of the nation's political leadership and economic elite invested in the crop. Especially during the one-party republic under Kamuzu Banda (1964-1994), the government provided increased support for large estates that specialized in producing burley and flue for export. As a result, tobacco land expansion, production and export grew steadily. In the late 1980s, Malawi adopted a policy of liberalization in line with demands from the World Bank and the International Monetary Fund (IMF): in the 1990s, smallholders could register to grow burley (not flue), and state-managed tobacco trade institutions such as KFCTA, one of the major national players in trade, closed. In practice, the dismantling of state monopolies gave more influence to large foreign cigarette manufacturers and leaf trading companies over local

[3] For tobacco land use data, see the Foreign Agricultural Service of the US Department of Agriculture at http://www.fas.usda.gov/cots/tobacco.asp and the Food and Agriculture Organization of the United Nations at http://www.faostat.fao.org

smallholders. Until 2006, transnational tobacco companies have owned all tobacco processing plants in Malawi (Kapito 2000). Only in 2006, did government-run Auction Holdings Ltd. (AH) establish the leaf processing company Malawi Leaf to end the near monopoly of foreign-owned tobacco processors in the country (Kakwesa 2006).

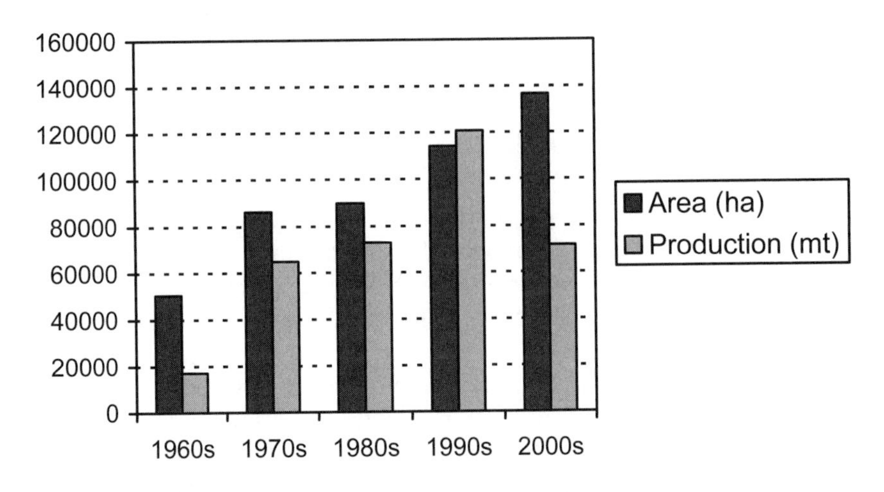

Fig. 14.1. Development of tobacco area (in hectares) and production (in metric tons) in Malawi, 1961-2006 (ten-year average) and six-year average for 2001-2006; FAOSTAT database (http://www.faostat.fao.org).

In the growing of flue and burley, estate owners and managers would have exclusively engaged family members, seasonal and migrant labor, and tenants (and still do so on 30-40 percent of the arable land). Since the mid-1990s, tobacco smallholders account for about 70 percent of the arable land and national tobacco production. In spite of declining production, tobacco land expansion continues well into the 2000s (Figure14.1), indicating, among others, that no great threats to estate farming have emerged such as the land acquisition movement in Zimbabwe in 2000 (Gossage 1997, Latner 1997, Rubert 2005).

Tobacco's rapid expansion is reflected in land acquisition by landlords ('tobacco barons') who increasingly seized large tracts of land in areas which hold most favourable soil and infrastructure endowments. For example, in Namwera (Mangochi District, bordering Mozambique), the area under estate tobacco in all cultivable land has increased drastically over one century from <1 percent to almost 60 percent (Table 14.2)

Table 14.1. Tobacco exports of Malawi by varieties as per May 1995, in kilograms

	Burley	Flue-cured	Dark Fire-cured	Sun-cured	Oriental	TOTAL
EU-25	5,071,825 57%	4,002,432 88%	470,016 100%	42,366 100%	554,057 100%	10,140,696 70%
Germany	1,581,910	211,300	10,000	200	357,247	2,160,657
United Kingdom	48,560	2,017,382	33,692	14,166	0	2,113,800
Czech Republic	1,516,000	0	0	0	0	1,516,000
Poland	204,960	814,400	100,000	0	196,810	1,316,170
Hungary	439,605	291,360	0	0	0	730,965
Denmark	376,130	264,360	0	28,000	0	668,490
Spain	501,120	9,600	63,000	0	0	573,720
Belgium	45,880	226,830	263,324	0	0	536,034
Italy	357,660	167,200	0	0	0	524,860
Other Countries	3,870,028 43%	522,250 12%	0 0%	0 0%	0 0%	4,392,278 30%
Turkey	1,201,720	0	0	0	0	1,201,720
USA	909,628	138,850	0	0	0	1,048,478
Russia	1,008,000	0	0	0	0	1,008,000
Philippines	611,760	0	0	0	0	611,760
South Africa	138,920	383,400	0	0	0	522,320
TOTAL	8,941,853	4,524,682	470,016	42,366	554,057	14,532,974

Source: TAMA Country Statement (24 October 1995)

From 1974 to 1982, a tobacco price boom and large institutional support by the Banda regime attracted many growers. Land seizure went hand in hand with the activities of international donors such as the African Development Bank and the German Agency for Technical Co-operation, funding and implementing integrated rural development projects (e.g., bridges, tarmac roads). These projects were meant for smallholder development but in practice served the profits and vested interests of the 'tobacco barons.' Finally, with the liberaliszation of the burley market in the 1990s, much more growers have taken up (smaller) leaseholdings (Table 14.2).

Table 14.2. Growth of leasehold tobacco estates in the tobacco growing zone of Namwera, Mangochi District (1910-1995)

	Number	Hectares	Percent of NAL[2]
1910	NA[1]	40	0.1
1937	NA[1]	423	0.5
1974	ca. 40	9,300	10.4
1982	117	38,510	43.2
1985	NA[1]	39,849	44.7
1990	NA[1]	45,000	50.5
1995	359	50,207	56.4

[1] Data not available.
[2] Net Agricultural Land, defined as total land area minus physical and legal limitations, ~89,000 hectares.

In Namwera, as in the rest of the country, very little land is held as freehold estate according to which the owner has title to the land free and clear of encumbrances (freehold land is rarely acquired now and remained static since 1973). While smallholders using land under customary land tenure are normally allocated land by their community leaders, most of today's 400 estates in the Namwera area are leasehold estates, 'reserved land' under the colonial authority and now negotiated with the Government of Malawi, usually through traditional users and local leaders. In Namwera, white estate owners managed to take leashold land exclusively from natural woodlands ('chief's undeveloped land'). In addition to leaseholding, land markets have developed which were not known prior to British occupation. Today, commercial land transactions in Namwera involve the National Bank of Malawi (NBM), ADMARC and Spearhead/Malawi Young Pioneers (MYP), accounting for about 40 percent of all transactions carried out until 1995 (Table 14.3) (Bosworth et al. 1997, Latner 1997).

At present, tobacco contributes about 13 percent to the total gross domestic product and generates 23 percent of total tax revenues (Jaffe 2003), accounting for more than 70 percent of the country's export earnings (FAO 2003). Figures for tobacco-related employment range from 650,000 to 2 million out of a total workforce of 5 million (FAO 2003), and estimates show that 70 percent of the working population are indirectly employed in tobacco (Kapito 2000). Over 98 percent of Malawi's leaf tobacco is exported (van Liemt 2002), and tobacco also accounts for 9.2 percent of Malawi's imports (ITC 2006b) (global leaf buyers in Malawi import tobacco from Mozambique and Zambia, process the tobacco in Malawi, and re-export to global markets). Malawi has one of the lowest smoking rates in the world (12.7 percent of adults over 18 smoke) (Mackay et al. 2006). Cigarette companies of PM/Altria and BAT, through a handful member states of the European Union (EU), account for the bulk of imports from Malawi (with UK manufacturers absorbing nearly half of all Malawian flue exports) (Table 14.1).

Malawi, together with just a handful net exporters of tobacco in the world, could be expected to have high transition costs (e.g., employment losses) over the long term if global demand falls due to regulation measures exerted through FCTC (Jacobs et al. 2000), or if prices continue to fall as they have done throughout the 2000s. However, the tobacco industry exaggerates the employment losses and other economic implications of strong tobacco control measures in tobacco growing countries (AAEC 1993, Warner et al. 1996). More research is needed to understand corporate practices to obstruct crop diversification and alternative livelihoods for tobacco farmers, and the impact of ending Malawi's economic dependency on tobacco.

As acknowledged by Bingu was Mutharika, Malawi's new president since 2004, the long term global decline in tobacco raises the issue of alternative land use policies. When addressing issues such as the end of growing, crop diversification, land reforms, or alternative (non) rural livelihoods, powerful key actors and their strategies in a complex interlinking network of local to global players are to be considered. Currently, remnants of a colonial (para)statal fix such as TAMA, TEAM, and ADMARC exist alongside strong private organizations such as buying companies/traders, commercial graders, processing plant owners, auction holders, and transporters.

Table 14.3 Types and actors of transactions of tobacco estate land in Namwera, eastern Mangochi District

	Malawian (African) Estate owner (N_1=91)			Malawian (white settler) Estate owner (N_2=22)			TOTAL (row) (N=113)	
	N	ha	%	N	ha	%	ha	%
1. Taken from customary land	67	8,746	59.7	17	4,165	42.9	12,911	53.0
a. cultivated smallholder land	(19)	(690)	(4.7)	(0)	(0)	(-)	(690)	(2.8)
b. chief's undeveloped land[a]	(48)	(8,056)	(55.0)	(17)	(4,165)	(42.9)	(12,221)	(50.2)
2. Purchased from …	16	4,337	29.6	7	4,861	50.0	9,198	37.7
a. individuals	(10)	(2,390)	(16.3)	(6)	(2,324)	(23.9)	(4,714)	(19.3)
b. institutions[b]	(6)	(1,947)	(13.3)	(1)	(2,537)	(26.1)	(4,484)	(18.4)
3. Herited from …	8	1,274	8.7	1	162	1.7	1,436	5.9%
a. family members	(7)	(1,235)	(8.4)	(1)	(162)	(1.7)	(1,397)	(5.7)
b. relatives	(1)	(39)	(0.3)	(0)	(0)	(-)	(39)	(0.2)
4. Otherwise acquired through …	2	298	2.0	2	520	5.4	818	3.4
a. trustee (bank)	(1)	(278)	(<0.1)	(1)	(502)	(5.2)	(780)	(3.2)
b. intermediate rent (individual)	(1)	(20)	(<0.1)	(1)	(18)	(0.2)	(38)	(0.2)
TOTAL (column) (N=113)		14,655	100.0%		9,708	100.0%	24,363	100.0

Source: Structured interviews in 1996 among 113 estates (32% of all registered estates in 1995; H. Geist).

[a] Agricultural reserve land, i.e., fallow, (semi)natural forest/woodland/grassland.
[b] Banks, companies and (para)statal groups (e.g., National Bank of Malawi, Spearhead/Malawi Young Pioneers, ADMARC.

Established in early 20th century, TAMA and the MoA's Tobacco Exporters' Association of Malawi (TEAM) still operate to protect and develop the tobacco industry and its export interests, as do various other (colonial) state and parastatal tobacco organizations such as ADMARC and the MoA's Tobacco Control Commission (TCC). Thus, the tobacco business remains a heavily regulated sector (e.g., farmers must register to grow flue on estates, flue must be sold over the auction floors, and permits are required to grow burley in smallholder clubs) (Latner 1997, Tobin and Knausenberger 1998).

Three auction houses in Blantyre-Limbe (south), Lilongwe (central) and Mzuzu (north) so far provide the main markets for estate-grown tobacco. Since 1962, the auction houses are under the umbrella of AHL (with ADMARC, TCC and TEAM as board members). AHL asks for a membership fee of US$100,000 so that most of the aspiring Malawian entrepreneurs are denied from participating (Houtkamp and van der Laan 1993). For the past half decade, prices realiszed from tobacco on the auction floors have been low. In 2005, government officials and an international lawyer have accused the leaf buying companies in Malawi of price-fixing and monopolistic buying practices (World Bank 2005, Stanbrook 2005).

Two US-based leaf buying companies, the world's largest tobacco traders – Limbe Leaf (Universal Leaf Tobacco Corporation, Virginia) and 'Alliance One' consisting of Stancom (Standard Commercial Corporation, North Carolina) and Dimon (Dimon Corporation, Virginia) – buy more than 90 percent of Malawi's tobacco. They undermine the economic power of local tobacco farmers and address, for example, landless burley tenants directly now to sell their produce to intermediate traders (instead of selling to the estate owner). As much as 40 percent of national tobacco estate production was marketed this way in the late 1990s (Latner 1997, Kapito 2000).

Since the innovations leading to the global rise of flue for use in global brands (such as 'Marlboro') were US-led, the global tobacco industry has become dominated by a handful of US firms (e.g., Dimon, Stancom, Universal, PM /Altria, BAT). By the early 1980s, the industry had been subsumed into a group of transnational conglomerates. By the late 1980s, it appeared that, due to tobacco industry re-organization and consolidation, and growing public concern of the health effects of smoking, at least some firms were entering a period of decline, visible through the move into diversified operations (such as food and finance). However, the Washington Consensus of the 1990s (World Bank, IMF) – i.e., the move towards deregulation of state-controlled markets and the liberaliszation of international trade – have opened a range of new opportunities for the tobacco transnationals, focusing their business solely on tobacco again (Kapito 2000, Cox 2003).

TOBACCO'S IMPRINT ON FOREST RESOURCES

The natural resource base of Malawi is part of a much wider geoecosystem called *miombo*. Featuring a hot, seasonally wet climate, this ecological zone extends from Tanzania and Congo in the north, through Zambia, Malawi and eastern Angola, to Zimbabwe and Mozambique in the south. Miombo is a vernacular word that has been adopted by ecologists to describe woody vegetation that is dominated by trees in the genera *Brachystegia*, *Julbernardia* and *Isoberlinia*, forming the largest more or less contiguous block of deciduous tropical woodlands and dry forests in the world (Millington et al. 1994). Wood from these trees provides an excellent (solid enough) and cheap ('free good') source of energy for flue- and fire-curing of tobacco. Since the mid-1980s, 90 percent of tobacco production in continental Africa has become located in countries covered by *miombo* woodlands (Geist 2006).

Empirical data on tobacco's imprint on forest resources in Malawi cover the period of the most rapid expansion of tobacco in the 1970s/1980s. For example, the total demand for wood by the tobacco industry was estimated to be one fourth of the total national demand in 1986 (Misana et al. 1996). However, no reliable nationwide land-cover change information exists for burley land expansion into the 2000s, except for anectodal evidence (e.g., Tobin and Knausenberger 1998) or some local data (e.g., Hudak and Wessman 2000).

In the 1972-1991 period, when the area under tobacco increased from ca. 54,000 to 117,000 ha and production more than tripled from ca. 30,000 to 113,000 mt, it was found, on the basis of remotely sensed data, that national forest cover declined drastically from 45 to 25 percent of the land surface, and that *miombo* species in areas most suitable for land clearance and mechanized large-scale estate farming were predominantly affected. For example, in Namwera where large flue and burley estates concentrate close to forest reserves and the international border, *Brachystegia* in flat zones was drastically reduced from 191,000 to 29,000 ha within 20 years (GoM/MoFNR 1993). A nation-wide survey implied that about half of the tobacco estates' total requirement of wood in the 1995/96 growing season was obtained from customary land outside the estates, concluding that "the net deforestation caused by the tobacco estates outside their boundaries is likely to be around 10,000 hectares per year" (Gossage 1997, 31). In Namwera, only 3 percent of all farms (totalling ca. 8,000 in 1995) – flue farms on large tracts of leasehold land ranging between 200 and 2,500 ha – consume more than 80 percent of wood used in the area, mostly taken from natural woodlands on state or customary land, or illegaly brought in from neighbouring Mozambique.

Commercial production of flue was first introduced in the Shire Highlands, spreading out from the Blantyre area, in1889. Transnational corporations and local business interests were encouraged to open large-scale tobacco estates in the hitherto sparsely populated, but rather fertile region. Soon, the rapid growth of area under tobacco created land scarcity among smallholders and – together with urban extension and the influx of refugees from Mozambique – built up population pressure and poverty conditions on customary land, further aggravating the removal of vegetation cover (Kalipeni and Feder 1999, Hudak and Wessmann 2000). In the late 1930s, when a tobacco auction was established in Blantyre-Limbe, several hundred square miles of forest land had already been converted to tobacco. More than 100 years later, the southern region, once the core agricultural area of Nyasaland, accounts for just about 20 percent of the total national tobacco production, most of the now heavily populated rural areas suffer from exhausted carrying capacities (a food security programme was even initiated), and most remaining biomass resources exist in areas where public access is restricted, i.e., game reserves, forest reserves, plantations, and traditional clan graveyards.

Throughout the early 20th century, tobacco production moved farther north, first into Lilongwe and later into Kasungu areas. By 1937, more than 400 square miles of land had been cleared of trees in both the southern and central regions to support the tobacco industry (McCracken 1983). In the late 1970s, a tobacco auction was established in Lilongwe, and in 1995, the central region alone produced almost 70 percent of the country's flue and burley (Latner 1997). However, limits of tobacco land expansion have been reached in certain areas where woody vegetation on customary land have been removed, and where tobacco estates have been established close to forest reserves, national parks, and game reserves (Gossage 1997).

A new auction house operating in Mzuzu since the 1990s reflects the fact that the latest and now final move of the tobacco frontier has reached the heavily forested areas with low population densities in the north (newly established tobacco auctions signal that a sufficient amount of tobacco for the world market is grown locally). In the mid-1990s, the northern region produced a little more than 10 percent of all flue and burley in Malawi which appears to be a low amount. Nonetheless, it demonstrates that "…tobacco is now heavily cultivated throughout Malawi" (Latner 1997).

CONCLUSION

Malawi is a prime example of how cigarette tobacco fundamentally transformed the nation's environment, economy and society in the course of a century. Contrary to abundant evidence of tobacco's impact on land degradation (notably deforestation through firewood for curing), the tobacco industry and its agricultural lobby continue to assert that tobacco's relative importance is minimal. This implies that natural resources such as woody vegetation dominated by *miombo* trees (generating, when burnt as 'free good,' excellent heat and fine aroma to cured tobacco leaves) continue to be depleted. This goes in parallel with the claimed economic benefits of 'green gold'. However, tobacco farming has improved the well-being of just a few individuals (European settlers, members of government, military, secret service and police) and a few organizations (e.g., auction holding, banks, insurance companies, transport businesses, agricultural marketing boards) at the expense of the majority of the Malawian population. It is for this reason, as well as the tobacco industries' influence on the government and economy of Malawi, that these individuals and institutions vehemently oppose efforts to regulate tobacco use, marketing and production through the implementation of national legislation and regulatory measures such as articles 17 and 18 of the FCTC. Economic benefits have not yet trickled down to smallholder farmers, tenants and the masses of seasonal workers leaving them as impoverished as before, or even poorer. As a matter of fact, Malawi is the thirteenth poorest country in the world (UNDP 2006)

REFERENCES

Arthur Andersen Economic Consulting (AAEC) (1993) Tobacco industry employment: A review of the Price Waterhouse economic impact report and Tobacco Institute estimates of economic losses from increasing Federal excise tax

Bosworth JL, Steele RJG, Mapemba LD (1997) The organisation, management and population of tobacco estates in Malawi. The report of the socio-economic survey 1996, Lilongwe, Malawi

Clay J (2004) World agriculture and the environment. A commodity-by-commodity guide to impacts and practices, Island Press, Washington DC

Cox H (2000) The global cigarette: Origins and evolution of British American Tobacco, 1880-1945. Oxford University Press, Oxford

Cox H (2003) Tobacco Industry. In: Mokyr J (ed) The Oxford encyclopedia of economic history, Oxford University Press, New York Vol. 5 pp 124-126

Food and Agriculture Organization of the United Nations (FAO) (1998). FAO statement on multisectoral collaboration on tobacco or health for the ECOSOC substantive session, FAO, Rome

---------- (2003) Issues in the global tobacco economy: Selected case studies. FAO Rome, ftp://ftp.fao.org/docrep/fao/006/y4997E/y4997e01.pdf.

---------- (2006) Major food and agricultural commodities and producers. Tobacco leaves, FAO, Rome, http:www.fao.org/es/ess/top/commodity.html?lang=en &item=826&year=2005

Geist H (2006) Tobacco. In: Ibd (ed) Our earth's changing land: An encyclopedia of land -use and land-cover change. Greenwood Press: Westport, CT, London, Vol. 2 pp 592-597

Government of Malawi/Ministry of Forestry and Natural Resources (GoM/MoFNR) (1993) Forest resources mapping and biomass assessment for Malawi GoM/MoFNR: Lilongwe, Malawi, and Satelitbild: Kinema, Sweden

Goodland RJA, Watson C, Ledec G (1984) Environmental management in tropical agriculture. Westview Press, Boulder, CO

Goodman J (1993) Tobacco in history: The cultures of dependence. Routledge, London, New York.

Gossage SJ (1997) Land use on the tobacco estates of Malawi. Report of the land use survey of tobacco estates in Malawi 1996, Lilongwe, Malawi

Houtkamp JA, van der Laan HL (1993) Commodity auctions in tropical Africa: A survey of the African tea, tobacco and coffee auctions, ASC Research Report 54
African Studies Centre, Leiden, NL

Hudak AT, Wessman CA (2000) Deforestation in Mwanza District, Malawi, from 1981 to 1992, as determined from Landsat MSS imagery. Applied Geography 20(2):155-175

International Tobacco Growers Association (ITGA) (1992) The economic significance of tobacco growing in central and southern Africa. Bardwell's: Harare, Zimbabwe

---------- (1995) Tobacco and the environment (Tobacco Briefing 4/95). ITGA, East Grinstead, UK

International Trade Center (ITC) (2006a) International Trade Statistics. United Nations Conference on Trade and Development, World Trade Organization, http://www.intracen.org/countries

--------- (2006b) Imports of Malawi. United Nations Conference on Trade and Development, World Trade Organization, http: //www.intracen.org/countries.
Jacobs R, Gale HF, Capehart TC, Zhang P, Jha P (2000) The supply-side effects of tobacco-control policies. In: Jha P, Chaloupka FJ (eds) Tobacco control in developing countries. Oxford University Press, New York, pp 311-341
Jaffe S (2003) Malawi's tobacco sector: Standing on one strong leg is better than on none. World Bank, Washington DC. http://www.worldbank.org/afr/wps/wp55.pdf
Kadzandira J, Phiri H, Zakeyo B, (2004) The perceptions and views of smallholder tobacco farmers on the state of play in the tobacco sector. Final report, Lilongwe, Malawi
Kakwesa T (2006) Malawi Leaf out to break cartel. The Nation Online [Malawi]
Kalipeni E, Feder D (1999) A political ecology perspective on environmental change in Malawi with the Blantyre Fuelwood Project Area as a case study. Politics and the Life Sciences 18(1):37-54
Kapito J (2000) The economics and politics of tobacco growing in Malawi. Proceedings of the World Conference on Tobacco or Health, American Medical Association, Chicago, IL
Koester U, Olney G, Mataya C, Chidzanji T (2004) Status and prospects of Malawi's tobacco industry: A value chain analysis. Ministry of Agriculture, Lilongwe, Malawi
Lambin EF, Geist H (eds) Land-use and land-cover change: Local processes and global impacts. Springer, Berlin, Heidelberg, GE
Latner K (1997) Market profile: Malawi (FAS Tobacco Circular 9702). US Department of Agriculture, Washington DC
Mackay J, Eriksen M, Shafey O (2006) The tobacco atlas. American Cancer Society, 2nd ed. Atlanta, Georgia
McCracken J (1983) Planters, peasants and the colonial state. The impact of the Native Tobacco Board in the central province of Malawi. Journal of Southern African Studies 9(2):172-192
Millington AC, Crtichley RW, Douglas TD, Ryan P (1994) Estimating woody biomass in sub-saharan Africa. World Bank, Washington DC
Misana SB, Mung'ong'o C, Mukamuri B (1996) Miombo woodlands in the wider context: Macro-economic and inter-sectoral influences. In: Campbell B (ed) The miombo in transition. Woodlands and welfare in Africa, Center for International Forstry Research, Bogor, pp 73-99
Otañez MG, Muggli ME, Hurt RD, Glantz SA (2006) Eliminating child labour in Malawi: a British American Tobacco corporate responsibility project to sidestep tobacco labour exploitation. Tobacco Control 15:224-230
Ponting C (1992) A green history of the world. Penguin Books, London
Reemtsma (1995) Tobacco: Driving force for economic development. Reemtsma Cigarettenfabriken, Hamburg, GE

Rubert SC (2005): Africa. In: Goodman J (ed) Tobacco in history and culture: An encyclopedia. Thomson Gale: Farmington Hills, MI, Vol. 1 pp 23-31
Stanbrook C (2005) Preliminary note on tobacco sales in Malawi. Washington DC, Brussels, Belgium
Tobin RJ, Knausenberger WI (1998) Dilemmas of development: Burley tobacco, the environment and economic growth in Malawi. Journal of Southern African Studies 24(2):405-424
United Nations Conference on Trade and Development (UNCTAD) (1995) Economic role of tobacco production and exports in countries depending on tobacco as a major source of income. Report 51627, UNCTAD, Geneva, CH
United Nations Human Development Programme (UNDP) (2006) Human development report 2006. hdr.undp.org/hdr2006/pdfs/report/HDR06-complete.pdf.
Universal Leaf Tobacco Company [Universal] (2006) World leaf production summary. http://www.universalcorp.com.
van Liemt G (2002) The world tobacco industry: Trends and prospects. International Labor Organization, Geneva, CH, www.ilo.org/public/english/dialogue/sector/papers/tobacco/wp179.pdf.
Warner K, Fulton G, Grimes D (1996) Employment implications of declining tobacco product sales for the regional economies of the United States. Journal of the American Medical Association 275(16):1241-46
World Bank (2005) Pathways to greatery efficiency and growth in the Malawi tobacco industry. A poverty and social impact analysis, Lilongwe, Malawi
Yach D, Bettcher D (2000) Globalisation of tobacco industry influence and new global responses. Tobacco Control 9:206-216

INDEX